Pitt Series in Policy and Institutional Studies

Steel, State, and Labor

Mobilization and Adjustment in France

Anthony Daley

University of Pittsburgh Press
Pittsburgh and London

Published by the University of Pittsburgh Press, Pittsburgh, Pa., 15260

Manufactured in the United States of America
Printed on acid-free paper

Library of Congress Cataloging-in-Publication Data

Daley, Anthony.
Steel, state, and labor : mobilization and adjustment in France / Anthony Daley.
p. cm. — (Pitt series in policy and institutional studies)
Includes bibliographical references and index.
ISBN 0-8229-3918-5
1. Steel industry and trade—France—History. I. Title. II. Series.
HD9522.5.D35 1996
338.4'7669142'0944—dc20 95-23147
CIP

A CIP catalogue record for this book is available from the British Library.
Eurospan, London

The author and publishers are grateful to the Fédération Française de l'Acier for permission to reproduce the map presented in figure 3.1, "La France Sidérurgique."

To my mother, Joan B. Daley, and to my father,
F. Edward Daley (1912–1982),
both of whom taught me the values of
perseverance and hard work

Contents

Tables and Figures

Tables

Figures

Abbreviations

AFPA	Association pour la Formation Professionnelle des Adultes (National Professional Training Center for Adults)
ANPE	Agence Nationale pour l'Emploi (National Employment Agency)
AS-FNE	Allocation Supplémentaire du Fonds National de l'Emploi (supplemental funds from the FNE)
ASSEDIC	Association pour l'Emploi dans l'Industrie et le Commerce (Local Unemployment Compensation Fund)
Assimilor	Association de la Sidérurgie et des Mines de Fer Lorraine
BPP	base-point pricing
CAA	*cession anticipée d'activité* (early retirement scheme)
CAPA	Caisse d'Amortissement pour l'Acier
CFAS	Compagnie Française des Aciers Spéciaux
CFC	*contrat formation-conversion* (conversion and training contract)
CFDT	Confédération Générale Démocratique du Travail
CFTC	Confédération Française des Travailleurs Chrétiens
CGC	Confédération Générale des Cadres
CGPS	Convention Générale de Protection Sociale
CGT	Confédération Générale du Travail
CGTU	Conféderation Générale du Travail Unifié

CIASI Comité Interministériel pour l'Aménagement des Structures Industrielles (Inter-Ministerial Committee for the Development of Economic Structures)

CICC Compagnie Industrielle Chiers-Châtillon

CLIF Compagnie Lorraine Industrielle et Financière

CNPF Confédération Nationale du Patronat Français

CODEFI Comité Départementaux d'Examen des Financements des Entreprises (Departmental Committee for the Examination of Private-Sector Finance)

CPS Convention de Protection Sociale

CSSF Chambre Syndicale de la Sidérurgie Française

DA *dispense d'activité* (exemption from work)

DATAR Délégation à l'Aménagement du Territoire et de l'Action Régionale (Department for Territorial Development and Regional Action)

DDE Direction Départementale de l'Emploi (Departmental Employment Office)

DE Délégation à l'Emploi (Employment Department within the Ministry of Labor)

DNEL Denain Nord-Est Longwy

ECSC European Coal and Steel Community

EMS European Monetary System

EMU Economic and Monetary Union

ETUC European Trade Union Confederation

Eurofer European Confederation of Iron and Steel Industries

FANE Société des Forges et Aciéries du Nord et de l'Est

FAS Fonds d'Assurance Formation (Training Insurance Fund)

FDES Fonds du Développement Économique et Social (Economic and Social Development Fund)

FF French francs

FGM	Fédération Générale de la Métallurgie (CFDT Metalworkers)
FGMM	Fédération Générale de la Métallurgie et des Mines (CFDT Metal and Mine Workers)
FNE	Fonds National de l'Emploi (National Employment Fund)
FO	Force Ouvrière
FTM	Fédération des Travailleurs de la Métallurgie (CGT Metalworkers)
GESIM	Groupement des Entreprises Sidérurgiques et Minières
GIS	Groupement des Industries Sidérurgiques
GRD	Garantie de Ressources pour les Demissions (Fund for the Guarantee of Income after Resignation)
GRL	Garantie de Ressources pour les Salariés Licenciés (Fund for the Guarantee of Income after Layoff)
INSEE	Institut National de la Statistique et des Etudes Économiques (National Institute for Statistics and Economic Studies)
IRSID	Institut de Recherche de la Sidérurgie (Steel Research Institute)
JOC	Jeunesse Ouvrière Chrétienne (Young Christian Workers)
MF	million francs
MoF	Ministry of Finance
MoI	Ministry of Industry
MRP	Mouvement Républicain Populaire (Popular Republican Movement)
MT	million metric tons
OECD	Organization for Economic Cooperation and Development
PCF	Parti Communiste Français (French Communist Party)
PED	European Development Pole
PS	Parti Socialiste (French Socialist Party—post-1969)

SAFE	Société des Aciers Fins de l'Est
SEA	Single European Act
SFIO	Section Française de l'Internationale Ouvrière (French Socialist Party pre-1969)
Sidélor	Union Sidérurgique Lorraine
SMK	Société Métallurgique de Knutange
SMN	Société Métallurgique de Normandie
SMS	Société Mosellane de Sidérurgie
SNAP	Société Nouvelle des Aciéries de Pompey
Socadev	Société d'Industrialisation—Normandie (Regional Development Company for Normandy
Sodilor	Société d'Industrialisation—Lorraine (Regional Development Company for the Lorraine)
Sodinor	Société d'Industrialisation—Nord (Regional Development Company for the Nord)
Sollac	Société Lorraine de Laminage Continu
Solmer	Société Lorraine et Méditerranéene de Laminage Continu
Somidev	Société d'Industrialisation—Méditerranée (Regional Development Company for the Mediterranean region)
SSE	*section syndicale d'entreprise* (union section at the workplace)
TUC	*travaux d'utilité collective*
UCPMI	Union des Consommateurs de Produits Métallurgiques et Industriels
UD	*union départementale* (departmental union)
UIMM	Union des Industries Métallurgiques et Minières
UL	*union locale* (local union)
UNEDIC	Union Nationale pour l'Emploi dans l'Industrie et le Commerce (National Unemployment Compensation Fund)

UNICE	Union des Confédérations de l'Industrie et des Employeurs de l'Europe (European Employers Association)
Usinor	Union Sidérurgique du Nord de la France
WFTU	World Federation of Trade Unions

Acknowledgments

In the course of writing this book, I incurred a number of institutional and personal debts. This work benefited extraordinarily from the specialized resources of various libraries and archives. It would not have been possible without the generosity and hard work of the staffs at the following libraries: the Institute for Industrial Relations, University of California, Berkeley; the Bibliothèque Nationale, Paris; the Chambre Syndicale de la Sidérurgie Française (CSSF), Paris, renamed the Fédération Française de l'Acier; the Bibliothèque Municipale, Metz; the library and documentation center of the IG Metall, Frankfurt; and the International Iron and Steel Institute, Brussels. I owe a special debt of appreciation to Chantal Granger on the staff of the documentation services of the CSSF.

I discovered rich empirical materials in the Archives Départementales de Moselle (Metz), Archives Départementales de Meurthe-et-Moselle (Nancy), and the archives of the Fédération Générale de la Métallurgie et des Mines—CFDT. The staffs were particularly helpful in securing access to materials that are usually difficult to find. In particular, I am indebted to Blandine Landas of the FGMM-CFDT for her help in obtaining complete and unhindered access to the accumulated archival materials of the Federation.

This work could not have been done without the contributions of dozens who participated in the process of industrial transformation. Many thanks to officials at the French Ministry of Industry, Sacilor, Usinor, the CSSF, Groupement des Entreprises Sidérurgiques et Minières, the Wirtschaftsvereinigung Eisen und Stahl, the European Communities, and the International Iron and Steel Institute for their conversations and useful explications. In particular, Ruprecht Vondron, secretary general of the WVES, and Jacques Ferry, former president of the CSSF, spent numerous hours explaining current and previous steel policy. Likewise, I extend my gratitude to David Brew and Peter Squarttini of the European Communities and Wenceslas Baudrillard and M. Cornier of the Ministry of Industry. Finally, Gérard Pressensé and Henri-Claude Bonnotte of Sollac-Fos, as well as Pierre Amet and his staff of the External Relations Department of Sollac-Florange answered questions, facilitated interviews, and arranged plant tours.

European steelworkers have suffered the trauma of job displacement. Dozens of French steelworkers and union officials at all levels were extraordinarily helpful in discussing local and national struggles for steel employment. I owe a special thanks to the following: Robert Chatelain, Jacky Chevalier, Albin Cinus, Robert Hautcolas, Bernard Lamirand, M. Lassagne, Gérard Lopparelli, Michel Martin, Xavier Phan Dinh, Belmire Rodriguez, Bernard Tomasina, Massimo Trignoli, Claudie Trzeciack, Maurice Vuillrume, Raymond Zimmerman of the CGT; André Bauduin, Michel Bruniau, Bruno Creton, Robert Giovanardi, Marc Gotté, Marcel Grignard, Gervais Hans, Joel Holin, François Introvigne, Vincent Keller, Raymond Laffineau, Michel Marti, Walter Paini, Jean-Pierre Querquelin, Pierre Robert, Jean Rondeau, Richard Stawirski of the CFDT; and Gilbert Fuchs, Charles Haury, and Henri Mallet of the FO.

I especially appreciate the help offered by Odette Hardy-Hémery, who clarified the unique problems of the steel industry in the Nord. François Introvigne facilitated numerous interviews in the Lorraine, warmly extended his hospitality, and engaged in hours of debate over the future of Lorraine.

This book began as a doctoral thesis at the University of California, Berkeley. As the chair of my dissertation committee, John Zysman provided the right mix of carrots and sticks to motivate this perfectionist. Jann Matlock contributed affection and intellectual feedback at the early stages of this project.

This manuscript was dramatically transformed in the following years, as I explored union organizations and labor market policy. I presented pieces of this work at various conferences. The following individuals contributed comments to various parts of it: Christopher A. Allen, Robin Gaster, Miriam Golden, Peter Hall, Chris Howell, Mark Kesselman, Peter Lange, Steve Lewis, Jonah Levy, Andy Markovits, Jonas Pontusson, Martin Schain, Serenella Sferza, Rand Smith, Sidney Tarrow, Kathleen Thelen, Joel Wolfe, and Sharon Zukin.

This study has benefited from other sources as well. In financial terms, it could not have been completed without funding from the Council for European Studies, the French-American Foundation, and Wesleyan University. I am especially grateful to the latter for a supportive environment. Many thanks to the staff of the Government Department—Lee Messina and especially Janet DeMicco—for lending their skills and ensuring that I fulfilled my other institutional responsibilities while I completed the manuscript.

I reserve a special thanks for David Cameron who read the penultimate manuscript once and several chapters of the final draft twice. He offered detailed and constructive criticism, and he maintained staunch support for the final project.

I have learned much from the work of George Ross over the years, and his insights into French politics and political economy inspired this research. He commented several times on the manuscript both in the form of conference papers and as a finished product. He seemed to know what I was trying to argue even before I did, and I am very grateful for his critical feedback.

Nancy W. Gallagher made this book possible. She read and critiqued numerous drafts and pushed me to clarify my arguments, provide appropriate evidence, and make this manuscript accessible to readers outside the narrow field of French politics. Most important, she continued a courtship even when my thoughts drifted to individual chapters.

Steel, State, and Labor

ONE

Industrial Politics in France

The creation of wealth depends on the capacity of economic actors to adapt to market signals. Such adaptation, in turn, poses fundamental questions about the distribution of resources. This book investigates the interaction among business, labor, and the state in France in the second half of the twentieth century, and it reveals how political dynamics refract market pressures. It explores in detail the process of industrial adjustment—defined here as company response to market signals—in the French steel industry. French steel companies struggled in the 1950s, flirted with bankruptcy in the 1960s and 1970s, and became profitable only in the 1980s. That achievement required fundamental changes in the traditional French model of industrial adjustment. Changes in the decision-making process itself, rather than market signals alone, explain why profitability came at the expense of jobs, vast amounts of state aid, and less national control over industrial decision making.

Successful adjustment in the French steel industry occurred only after tensions involving business, labor, and the state had been resolved. Those tensions resulted from competing and frequently conflicting pressures. Each party's ability to secure its goals depended on its influence over the other two. Until the 1980s, these tensions created a mismatch between product and labor market policies that undermined effective industrial adjustment. Ironically, the provisional resolution of the steel stalemate came too late to satisfy the original goals of any of the key players.

Steel managers in postwar France wanted to maintain their way of life despite a hostile international environment, a threatening state, and a radical transformative labor movement. They managed to shape the European Coal and Steel Community (ECSC) so as to cartelize the domestic market, but such protectionist strategies only postponed the pressure to become more competitive. The industry desperately sought managerial independence from state encroachment, yet lacked the market position to remain solvent without subsidies. The managers convinced state modernizers to exclude labor from policy making, but these paternalistic attitudes sparked disruptive protests and slowed the spread of modern production techniques.

State planners sought to use industrial policy both to modernize the French economy and to increase support for the regime. They hoped that a more efficient steel industry would be an engine of growth and stimulate a mass-production cycle in which better productivity brought wage increases and increased demand for other French products. Yet the need for cheap steel obliged the state to compensate for price controls by subsidizing steel production. To woo the industrialists, the state financed new facilities without insisting that inefficient plants be closed. They also gave managers a free hand with labor in return for cooperation on product market decisions. Labor protests against paternalistic exclusion, however, threatened government stability and forced the state to guarantee protection for unproductive jobs.

Labor wanted to participate in policy decisions that affected the terms of employment, but the unions were not organized well enough to demand direct involvement. Therefore, they sought indirect influence by encouraging state intervention. Because the unions had leverage only when social peace seemed to be threatened, protest became a regular part of labor politics. The government tried to avoid social unrest through crude strategies designed to buy short-term acquiescence, yet it lacked the sophisticated tools needed to bring labor into the adjustment process. Conflictual unionism required constant mobilization that was hard to sustain as hard times and technological change decimated the ranks of blue-collar workers. Further, confrontational politics often prevented labor from being a responsible bargaining partner when management was ready to make authority relations more inclusive.

The stalemate started to unravel when the state nationalized the steel industry, sidestepped both unions and managers, shut down aging facilities, and focused on the most sophisticated technology. The strategem succeeded, in part, because the new generation of industrial managers was more attuned to market dynamics, less concerned with control issues, and less wedded to the owners. A changing skill base and inter-union struggles had produced a more accommodating labor movement. Policy makers continued to buy off opposition, but also developed more sophisticated strategies, such as creating new jobs in the steel regions and encouraging some union administration of worker conversion programs. Most important, state officials decided to stay within an ECSC whose cartel had collapsed due to growing competition. Embracing Europe and adopting orthodox economic policies made the state less willing to placate labor at the cost of productivity and more willing to give industry back the freedom and flexibility it needed to compete. Thus, successful adjustment could occur only after the steel industry, labor, and the state altered their interests and adopted new strategies for influencing industrial politics.

Three broad themes guide this book. First, industrial adjustment is not a purely market phenomenon. International markets send signals that are refracted by domestic firms, labor unions and other organized interests, and state bureaucrats. Second, strategies, assets, and vulnerabilities change over time as actors learn to play the game more effectively and as shifts in technology, markets, and political coalitions alter the nature of the game being played. Third, victories or defeats can be short-lived in a fluctuating environment. Put another way, industrial adjustment is a perpetual struggle.

The Politics of Industrial Governance

Viewing industrial adjustment in France as an ongoing bargain forged among actors with different assets, vulnerabilities, and strategies enables one to reassess dominant analyses of French politics. It suggests that stereotypes about the French—either in their own terms or in comparative perspective—have led to unhelpful generalizations. This is an effort to rethink how labor, business, and the French state interact and to understand how actors adapt to changing circumstances.

Steel has been among the most studied economic sectors in France because it epitomizes the challenges of industrial policy. In the 1950s, scholars used the steel sector's resistance to European integration as evidence of the "backwardness" of French business.[1] In the 1960s, observers depicted the steel industry as a paradigm of the planning process in which government agencies channel investment and direct industrial adjustment.[2] In the early 1970s, steel seemed the industry in which the state's capacity for selective intervention was most advanced.[3] By the 1980s, the industry was viewed as the quintessential lame duck that absorbed huge state subsidies.[4] In the 1990s, the CEO of the steel giant Usinor-Sacilor graced the covers of slick business magazines as the embodiment of the new manager.

Such characterizations exaggerate discontinuities and miss the connections linking various stages in the evolution of the French steel industry. These snapshots capture neither the weight of historical choices nor the interplay of company strategy, labor mobilization, and state intervention. The analysis that follows details patterns of interaction and explains a sequence of baffling choices for two generations of beleaguered executives, frustrated policy makers, and angry workers. In particular, it shows how strategies of individual actors are constrained by past choices and how they develop new strategies in response to those constraints.

In general, studies of French industrial adjustment typically examine the initiatives and constraints for capital, labor, or the state in isolation.[5] Analysts of French industrial policy stress the policy tools of the strong state.[6] They see conservative capitalists and radical unions as annoyances that hinder rational planning and force leaders to spend additional resources to fend off attacks or appease aggrieved constituents. Specialists in industrial relations focus on social protest as an indicator of labor's weakness and a sign that unions have little constructive role in negotiating and implementing adjustment policy.[7] Economic experts emphasize how government interference and labor protests have compromised market efficiency.[8]

Industrial adjustment is more than the sum of its parts. Deconstructing adjustment politics by considering the perspective of one actor at a time can be just as misleading as generalizing about industrial governance from the circumstances of a single sector at a particular moment. Observers tend to adopt uncritically one actor's assumptions. Likewise, they tend to distort the power, coherence, and choices of other actors. A single-actor perspective analyzes adjustment by maximizing only one set of values.

Adjustment strategies are shaped by patterns of state-society relations, interest representation, and social mobilization. Thus, the steel case sheds light not only on the process of industrial adjustment, but also on larger questions about French politics. Analyzing the steel industry in terms of a trilateral adjustment process calls into question some of the dominant generalizations about French politics and provides new insights into longstanding scholarly debates.

State-Society Relations

French public policy reveals an acute tension between the desire to centralize power and thus allocate resources effectively and the need to include a wide array of interests—either to buy their support, increase productivity, or deepen democratic politics. Proponents of the strong state in France have long squared off against those seeking to democratize the state. This debate dates from struggles between center and periphery under the absolute monarchy of the seventeenth century.[9]

Arguments about the appropriate role of the state in France cut across traditional political cleavages. In the Fifth Republic, the "strong statists" included "historic" Gaullists who glorified the nation, as well as Communists and Left Socialists who sought to reinforce planning at the center. The "weak statists" ranged from those on the right seeking to decentralize state administration and empower localities to those on the left who accepted a limitation of state power and the utility of market interactions. Because both strong and weak statists could carry the interests of labor, it is not surprising to find within the labor movement a deep ambivalence about the role of the state.

Most students of French politics have concluded that—with all the usual qualifiers—strong statists eventually prevailed. Ever since Alexis de Tocqueville complained about the weakness of associational life, France is said to be exceptional because of the strength of the state and its ability to structure opportunities for social interests.[10] In his pathbreaking work on authority relations within organizations in France, Michel Crozier describes a "fear of face-to-face" relations.[11] He extended this analysis to French society in general and described it as "blocked"—rigid, hierarchical, poorly adaptive, and highly ideologically driven.[12] Stanley Hoffmann argues that before 1945 stagnant social groups with vested interests in preindustrial values—the *haute bourgeoisie*, the lower middle classes, and the peasants—conspired to resist the "machine age" and its disruptive effects on social norms and social classes.[13] This lock on social development was not completely broken at Liberation. Instead, the confrontation between the "silent (but fragmented) majority" and the numerically inferior working classes led to "nonpragmatism" and "nonreformism" on both sides.

Society was "stalemated."[14] Given the high level of intergroup distrust, social actors preferred to have the state resolve their dilemmas because it was less obviously biased. Protest was the mode of expression in the absence of coherently organized interests.[15]

Many analyses of French social groups focus on internal processes. French associational life is frequently characterized as ideational and hierarchical. French groups are alleged to be undemocratic and driven by ideological norms more than practical needs or member services. The evidence varies depending on the issue. Education in France, writes John Ambler, is perpetually politicized.[16] In contrast, agricultural politics fluctuates between practical cooperation and politicization.[17] Subnational cases show variation in what Robert Salisbury calls "expressive" benefits.[18] By focusing on internal group processes and membership benefits, however, these analyses tacitly accept the prevailing view of a state-driven society.

Yet the relationships among groups or between groups and the state remain sources of controversy. Under what conditions can groups influence the French state? Must influence remain internal or external to policy making? Must new actors be included explicitly at the agenda-forming stage? What are the mechanisms of inclusion and exclusion? How could French politics be made more open?[19] Given the weight of the French state, the popular assumption is that excluded groups can be included only when the state changes hands.

This study suggests that political power is less a function of inherent group strengths and is more tied to relations among actors. Consequently, "excluded" groups can influence adjustment strategies in nontraditional ways. This implies that the French state is not as strong, French labor not as weak, and French capital not as constrained as scholars have argued.

Interest Representation in France

In a democracy, citizens should be able to pursue their interests and therefore influence the content of public policy. The sine qua non of democracy, of course, is the citizen's ability to vote for candidates expressing similar perspectives. Robert Dahl suggests that democratic governance (which he labels polyarchy) also depends on meaningful participation, an enlightened understanding of crucial issues, and some popular control over dominant political agendas.[20]

The relationship between the French state and the business community seems to contradict those principles. When in 1965 Andrew Shonfield described the French planning process as a conspiracy between big business and "big officialdom," he captured the official wisdom about the French adjustment process.[21] The government had policy tools to intervene in the economy. Industrial adjustment consisted of the ability to exclude interests who might either disrupt the planning process or espouse contradictory ideas. Groups were poorly organized and political parties were splintered. This was a state that was autonomous from society.[22]

Organized labor was excluded more than other groups. Those who justified such exclusion did so on the basis of the unions' radical transformative ideology. In the 1950s and 1960s, the Confédération Générale du Travail (CGT) was by far the largest union in France, was closely related to the Communist Party (PCF) that was itself a pariah, and analyzed the world in class terms. It had little interest in cooperating in probusiness policies. Consequently, state leaders dealt with capital on a bilateral basis, and managers dealt with the unions as little as possible.

Frank Lee Wilson was the first Anglo-American scholar to explore systematically the relations between interest groups and the French state and to consider both within a comparative inquiry of interest aggregation.[23] From a series of interviews with interest group leaders in the early 1980s, he concluded that the dominant pattern of group action in France was pluralism—competitive groups with the state relatively distant—but with two "peculiarities."[24] First, he found, certain groups are excluded from the negotiating process, but who gets excluded depends on the regime. Second, groups frequently use various modes of protest to achieve their goals. According to Wilson, there is a lack of social consensus over the proper role of government and what he calls a "Rousseauian aloofness" of the state from particular interests.

John Keeler countered that the French state has used a variety of mechanisms—including recognition, subsidy, statutory bodies, and biased information—to structure interest representation.[25] Consequently, "the French case has featured a neocorporatist trend"—a tendency for collusive behavior among groups and a relatively close relationship between interests and the state. He qualified his conclusions, however, by acknowledging that "interest group dynamics in France have been unevenly corporatized."[26] Using Wilson's three sectors—agriculture, business, and labor—Keeler characterizes the first as highly corporatist, the second significantly corporatized, and the third least corporatist.

The less the inclusion, the greater the social protest to articulate interests.

What are we to make of this controversy? First, generalizing at the social level is not very useful in addressing questions of interest inclusion in France. Ironically, both Wilson and Keeler admit to considerable variation in relations among interests and between interests and the state. In particular, as Keeler argues in a longer work, this type of argument tends to obscure the changing patterns of interaction *over time*.[27] Second, focusing on leadership attitudes rather than organizational behavior ignores the possibility that they do not always correspond. Without investigating processes and outcomes, one cannot examine the conditions under which interests are included in decision making. Moreover, attitude surveys do not always reflect "issue scope"—the possibility that some decisions or issue areas are more consequential or delicate than others. Third, because of this difference in issue relevance, bargaining may be asymmetrical and one side might enjoy a disproportionate amount of information, leverage, or capacity. In a struggle over resources, various coalitions may use different tactics. Fourth, dividing the economic universe into agriculture, business, and labor is misleading. It overemphasizes the role of agriculture, which seems to display similar styles of policy making and interest mediation in widely different political economies. *Business* is a highly differentiated category, depending at least partially on the size of an industry as well as its position in domestic and international markets. Finally, unions are not producers and do not bargain along the same dimensions as industrial or agricultural interests.

The attempt to categorize interest politics has stumbled most over the role played by organized labor. Both Wilson and Keeler refer to labor as a separate "sector"—unaffected by relations among other interests or between those interests and the state. In contrast, Andrew Cox and Jack Hayward suggest that business-state coordination has come at the expense of labor's integration.[28] They argue that the organizational weakness and fragmentation of labor in France prevent sustained bargaining—either with business or with the state. French unions are "lean" organizations. A small proportion of the labor force is unionized into five competing confederations. As a result, labor finds itself on the outside, while individual firms collude with the state bureaucracy on individual policies—a "corporatism without labor."

Pierre Birnbaum pursues further the argument of labor's "weakness." Social actors seek to maintain their autonomy and do not negotiate with each other.[29] Development of the state at the expense of social intermediaries in the eighteenth century, however, gave the state the power to alter social relations. Therefore, social actors turn to the government to intervene in conflicts. Corporatist networks in France would necessarily entail the sacrifice of powers and prerogatives by the state. Birnbaum concludes that "societal corporatism is unthinkable to the extent that trade unions refuse integration into a state that itself objects to such a notion for threatening its specificity."[30] Collusive behavior is impossible to the extent that interests guard their autonomy and the state is willing to fill the void.

Pluralism seems to be a misleading concept in a political system with a state wielding the policy tools available to the French state. Competition for resources may exist among voluntary groups that are nominally free from state control, but such competition is meaningless when the state has policy tools sufficient to act on its own. Pluralism, as the term is generally used, implies not simply the competition of organizations, but the ability of these interest organizations to make a difference in formal policy making. According to Nelson Polsby, power in a pluralist society cannot be centralized.[31] In France, the state seems to have the policy tools in some arenas to intervene to restructure social relations. Consequently, the empirical question becomes what influences state policy makers.

Likewise, only a few economic sectors in France can be adequately described as neocorporatist. Peak-level interaction has taken place through advisory committees. Yet the weight of those committees tends to be light. The very language used by the French is instructive. The word *corporatisme* connotes group egotism, exclusion of outsiders, and group defense of previous gains and vested interests.[32] In contrast to its Anglo-Saxon cognate, it does not stress networks of interaction.[33] The word is profoundly derogatory; *corporatisme* conflicts with the general interest, represented by the state and public authority. The actual behavior of the French departs considerably from this normative dichotomy. Private parties rarely sacrifice their own "sectional" interests, yet the pursuit of such interests by a rival is often considered illegitimate. The result is an ambiguity of conduct in which the public and private spheres are confused.

Because national peak associations frequently juggle concerns among member associations organized by industry, they tend to be one step removed from the adjustment process. In contrast, subnational analysis can examine the effects of policy on market forces and vice versa. It highlights the relationship between markets and interest formation, aggregation, and articulation. This level of analysis permits the exploration of variations in relations between interests and between interest and the state. The concepts of *mesopluralism* or *mesocorporatism* would characterize politics at lower levels of the political system.[34]

Here research on policy communities can be of considerable use in exploring relations among interests and between interests and the state.[35] The authors in this tradition alert us to the compartmentalization of knowledge and the importance of epistemic communities. They suggest that channels of influence may be open or closed during the formulation and implementation of policy. The "mobilization of bias," E. E. Schattschneider notes, may affect one actor's ability to influence policy outcomes because "some issues are organized into politics while others are organized out."[36]

Critics have attacked the French state's exclusion of labor and its economic policies that systematically ignored working-class interests. Where there were no unions, foremen became minor dictators. The result was a weakly institutionalized system of industrial relations.[37] Neither unions nor employers had a vested interest in developing an intricate system of rules to guide collective bargaining, and as a result the latter remained unstable.

The perpetual marginalization of labor suggests that neither pluralism nor corporatism effectively captures the dynamics of adjustment. Organized labor has not been well represented in policy-making circles at decisive stages of policy formulation—even after the Socialists achieved power in 1981.[38] Yet unions have frequently mounted protests outside the ordinary political process. Labor has protested against the actions of a state which it thinks wields the tools of adjustment. Thus, the strong state has become the target of extrapolitical action. This suggests that the tight linkages between some groups and the state can be loosened under certain conditions.

Social Mobilization and Protest

French political life seems characterized by outbursts of protest in which aggrieved groups resort to unconventional tactics to obtain their ends. Truckers block highways when the Ministry of Transportation alters the rules on speeding violations. Farmers dump produce in city squares to protest damaging agricultural policies. Vintners hijack truckloads of cheap imported wine and discharge the contents. Parents of private school students bang pots and pans in support of state subsidies. Train conductors stop their trains to pressure the railways to raise their wages. Redundant steel workers occupy a regional television station to dramatize their plight.

The French have developed a mythology of protest. The public display of anger has had regime-threatening consequences—the Terror of 1792 on constitutional monarchy, the mobs of 1830 on the Bourbon restoration, the storming of the Tuileries in 1848 on the regime of Louis-Philippe. The Communards in 1871 and the students and workers of 1968 threatened but failed to usher in a new political order. The strikers of the Popular Front in 1936 and the cold war disturbances of 1947–1948 contested the legitimacy of the established order. Each generation, it seems, has had its set of protests in which normal politics no longer seems to work, conflict becomes overt, and the streets are alive with tumult. As Charles Tilly claimed, France is contentious.[39]

Yet protest is not uniquely French. Direct action in Germany, Japan, and the United States—just to suggest three cases—has been used to influence foreign policy, to block the installation of nuclear weapons, to protest airport extensions, to promote environmental policy, and to demand reproductive choice. Closer examination suggests that the French engage less in protest than do other nationalities. Charles Taylor and David Jodice break down protest into five distinct measures—demonstrations, political strikes, riots, armed attacks, and assassinations.[40] On the last three measures, which we might label "protest with violence," France ranked consistently below Italy and the United Kingdom (two European countries of equivalent populations) between 1948 and 1977, a period including the resolution of two colonial wars that brought violence to French territory. Even in the first two categories, France ranked behind the United Kingdom on demonstrations and Italy on political strikes. Yet the French and their observers see the polity as protest-laden.[41]

The French case is important not for the quantity of protest but for the uses to which protest is put in France. Let us begin with assumptions about interest organization. A group wields the "weapon of the weak" when its interests are not acknowledged or accommodated.[42] We assume that protest is not a form of social recreation and that its usage is either goal-directed or identity-creating.[43] If we can find instances in which direct action (or even the threat of mobilization) consistently alters policy preferences, then we can temper the view of a strong French state with the will, expertise, and policy tools to intervene effectively in society and the economy. If we can observe labor using protest to secure tangible gains, then we can modify the image of French labor as weak and excluded from the process of industrial adjustment. Finally, if we can uncover evidence of business responding substantively to protest, then we can understand better what strategies companies use to shape their labor policy.

Methodology and Itinerary

The political nature of adjustment requires an investigation of what we might call *industrial governance*—the decision-making authority and institutional arrangements that regulate patterns of interaction among economic actors.[44] We must ask questions that assess the relative power of different actors. How does each interpret market signals and then affect the state's agenda? Under what conditions can labor govern with business? When will labor protest modify policy options? Can "veto power" be transformed into agenda-setting capabilities? If policy is modified by protest, how do we evaluate bids for political inclusion? Likewise, we must explain variations between what actors get when they are included in the decision-making process and what is gained by outside pressure.

Sectoral analysis approaches the political reality of an industry (with all the empirical holes that implies) that is represented by a trade association, that sees itself as acting collectively, and that is viewed by public policy leaders as an entity.[45] Likewise, union federations see this industry as an integrated whole, for they attempt to remove from competition wages for similarly skilled workers. The story of French steel, however, must be told within the context of European integration, erosion of autonomous domestic policy making, institutional coordination within the state, party conflict, and the strategies of national union organizations. It enables us to capture the mechanics of firm-level adjustment within the context of broader political mobilization.

Case studies force us to think through the appropriate unit of analysis for exploring our dependent variable. The attributes of the steel industry serve as a lens through which we can analyze prevailing patterns and variations of industrial interaction. By investigating a single industry, we encounter nuts-and-bolts conflicts as the actors experienced them.[46] Sectoral analysis thus skirts the sampling problem inherent in studies of microlevel organization. It also avoids the overgeneralities of macro analysis.

However, a case study begs the question of whether our findings may be applied even to the French political economy, not to mention other national contexts. This is an inherent problem of sectoral analysis that cannot be ignored. A fundamental theme of this book, however, is that the complex relations among the state, business, and labor can have unanticipated consequences, and these can be understood only within a holistic framework that emphasizes complexity and ambiguity in relations among the state, business, and labor.[47] Likewise, a case study permits us to explore that complexity over time.

The adjustment story in steel is a strong case. Large state subsidies, business's reluctance to innovate, or union protest over the terms of adjustment in this particular industry can be found in other French sectors. In steel, all three have been evident. If we can explain outcomes here, we can shed light on the distinctive pattern of adjustment in France. The case study in steel illustrates the dilemmas of state-led modernization, the problems of company innovation, and a behavioral reflex by unions that can best be summed up as protest.

This study uses a variety of data sources to investigate the process of industrial adjustment in the French steel industry. In 1984–1985 and then in summer 1989 and summer 1990, I conducted over a hundred open-ended interviews with officials in Directorate-General III (Internal Market and Industrial Affairs) and Directorate-General V (Employment and Social Affairs) of the European Union; state planners in the French Ministries of Industry and Labor; officials in the Chambre Syndicale de la Sidérurgie Française (CSSF), the trade association for steel; managers in the steel companies Usinor and Sacilor and their subsidiaries; and union officials within the CGT, the Confédération Générale Démocratique du Travail (CFDT), and the Force Ouvrière (FO) at central offices, regional unions, and plant locals. I also visited and toured the major integrated steel plants. The use of state archives in the departments of Moselle and Meurthe-et-Moselle enabled me to shed new light on adjustment in the early postwar period. The CFDT metalworkers' union (the Fédération Générale de la Métallurgie et des Mines) generously opened all its private archives for the period between 1945 and 1980. I perused local and national periodicals published by both the unions and the employers. Likewise, I had access to reports for the trade association and individual companies.

Any journey requires a map. Chapter 2 explains the origins and tensions within what has come to be known as the French model. If adjustment means choosing strategies to confront market changes, then who chooses what strategy is an important political issue. The chapter explores the struggles over state-led adjustment in France in the immediate postwar period. Political compromise in the 1940s created a mismatch in tools available for industrial and labor policy decision making. The state could exercise national control over industrial policy only if it gave employers the freedom to formulate labor policy on their own. Labor exclusion combined with political mobilization to generate popular protest against state-led adjustment. The remainder of this study explains the shift in the French model.

Chapters 3 through 8 explore company behavior, state industrial policy, work organization, union mobilization, and labor market measures to ease the cost of adjustment in the steel industry. Chapter 3 examines industrial strategies in steel in the twenty years after World War II. Owners sought to create a modern industry in the 1950s while preserving the paternalistic authority relations of the nineteenth century. As a result, a patchwork of different plants and work organizations characterized the industry, which by the early 1960s was facing competitive challenges in international (especially European) markets. State actors had mobilized capital to help the firms to adjust but, because they accepted the industry's key arguments about the necessity to diversify products, they failed to ensure profitability.

Chapter 4 probes the state's failure to challenge the exclusion of labor by the steel employers. Unions used confrontational rhetoric both to break the pattern of paternalism and to influence the adjustment process in a Sisyphean struggle. Yet the labor movement needed to overcome its own internal divisions in order to press the employers on issues of salary and employee participation. Failing in their attempts to be included at the bargaining table, the unions appealed to the state for assistance.

Chapter 5 investigates state efforts to create national champions in steel. These would be one or two firms enjoying economies of scale that would dominate national markets and eventually become major players in international markets. Government policy failed to create the mass-production industry so desperately sought by policy makers. The trade association mediated business-state relations and hampered the state's efforts to seize the initiative in coordinating production. Investments were poorly coordinated and inefficient plants were maintained because the trade association could manage state policy makers. Because mass production was slow to emerge, owners could delay the coordination of collective bargaining so as to control labor policy until the mid-1960s. State intervention, however, altered the nature of collective bargaining, as employment and wage negotiations assumed a national, rather than regional, scope. Consequently, the national champion policy simultaneously combined industry assets and spurred labor mobilization.

Chapter 6 examines how industrial decline was negotiated after the precipitous decline in world demand for steel products in 1974. By the mid-1970s, the steel industry was in virtual bankruptcy. Because it was slow to modernize, the industry was at a serious competitive disadvantage compared to its international rivals. The state intervened forcefully by nationalizing the companies and firing the old management team. As the employment crisis deepened, the unions had difficulty in adapting to new competitive conditions, viewing problems through old lenses and focusing on who owned the assets rather than changes in productivity. The combination of decentralized militancy and centralized negotiations made it difficult for both the companies and the state to formulate a labor policy that would ensure competitiveness. The unions eventually won greater participation, but because it was so late in coming, it brought meager returns.

Chapter 7 looks at the competitive success of the 1980s. Years of losses encouraged the now state-owned steel companies to appeal to the European Community for market protection. The coordinated approach of European producers forced the French industry to downsize and focus production on selected product groups. Europeanization of the industry, modernized plants and equipment, and a reduced work force eventually facilitated a return to profitability. Modernization involved reorganizing both work and authority relations premised on employee participation. A combination of state intervention and weak unions enabled employers to experiment with new social policy at the plant level.

Chapter 8 highlights the contradictions of adjustment in the steel industry. For the companies, profitability came at the expense of managerial control. Nationalization was the first step toward competitive success in foreign and domestic product markets. Layoffs were avoided and new jobs were created, but the unions became increasingly entangled in the administrative tasks of adjustment. While by the 1980s the state had the policy tools to intervene in both product and labor markets, intervention was extremely expensive and only partially successful in shifting employees out of steel and into other forms of activity. Thus, the state had its policy levers and the unions had participation. Neither could be content with the final model of steel adjustment, however, because the processes they had so long coveted did not translate into desired outcomes.

Chapter 9 considers the impact of two major policy choices—European policy and the embrace of economic orthodoxy—on contemporary adjustment politics in France. As soon as the state had the mechanisms to intervene simultaneously in product and labor markets, it discovered that state-led adjustment was no longer so effective in directing growth because the arenas of economic activity had shifted from the national to the international and local levels. By the 1980s, business had regained significant freedom of action. European policy expanded markets, while state labor policy eased the costs of restructuring. The ultimate constraint on business activity, nationalization, was crippled by shifts in political coalitions. Intellectual fashions of the moderate Left and Right favored entrepreneurial initiative and flexibility over state ownership and economic redistribution. Ironically, labor received considerably more formal involvement in the adjustment process, but saw its influence decline because protest could no longer spur reaction. In general, the balance between state and society had shifted to the latter, the pattern of interest articulation had been made more permeable, and social mobilization had become less effective for organized labor. The conversion of economic interests had changed the nature of the adjustment process.

TWO

The Adjustment Triangle

Industrial adjustment in France evolved continually in the postwar period. A number of excellent studies have examined the economic basis for this performance.[1] This chapter examines instead the political process of industrial adjustment and explores the uneven development of the state-led adjustment model. The empirical material in chapters 3 through 8 shows how these policy tools and political struggles shaped industrial adjustment in the steel industry.

The elements of state intervention were assembled during the period of reconstruction when postwar planners borrowed from earlier state efforts and combined them with new ideas to reform the economy and society. They relied on financial inducements to encourage business to invest in particular directions. Labor market policy sought to compensate for the problems caused by weakly organized unions that were incapable of negotiating a set of rules to govern transactions. The slow consolidation of labor policy tools, however, gave the government only limited power to intervene effectively in labor markets.

The mismatch of product market and labor market tools in France retarded the development of a mass production economy. Late development of mass production, in turn, reinforced union opposition. Yet French unions were not simply motivated by hostility to capitalism. They were low on members, short of funds, unable to sustain long strikes, and vulnerable to employer whipsawing—dealing with whatever union was most willing to cooperate. The rules of representation, however, by providing constant elections for shop floor institutions, enabled unions to represent the vast majority of French workers. Representation without organization tended to divorce bargaining from militancy, thereby destabilizing labor politics.

The French state therefore had both institutional strengths and vulnerabilities. Policy tools to influence product and labor markets changed over the postwar period, along with the relative power of government ministries and France's position in the international economy. The state could get much less of what it wanted in labor policy because policy makers compromised with business over the mechanisms of industrial policy. This settlement, in turn, weakened the full impact of industrial modernization. As a result, the strong French state became vulnerable to protest.

Political Struggles over State-Led Adjustment

The centralized state has a long tradition in France—bolstered by seventeenth-century absolutism, Napoleonic territorial organization, and Third Republic republicanism. The centralized state, however, proved impotent in the face of the German blitzkrieg. The political struggles of the 1940s resembled those of a half century later: how to forge effective national strategies given an overwhelming international constraint. According to a widely shared view, the character of the state was responsible for the tragedy of 1940. Consensus over the need for change, however, did not reflect agreement on what state powers and what industrial strategies would preserve and promote France's power and welfare in the modern world. Political struggles developed over the types of changes needed and how social actors would be mobilized to support those changes.

Modernizers confronted traditionalists in 1940. The former wanted the new production methods that had succeeded in other countries. Traditionalists wanted a return to a mythic past through an imposed class harmony. This was symbolized by the fights between the industrial experts and ideologues like Charles Maurras during the Vichy regime, which sued for peace with the Nazis. Vichy was the first attempt to remake French politics and society.[2] Despite wartime exigencies and competing visions of renewal, proponents of the authoritarian National Revolution shared a moral contempt for the squabbling republicanism that had degenerated into administrative deadlock. While upholding the importance of property, they disliked economic liberalism because, they argued, it misallocated resources. They feared the anomic effects of mass society and preached class harmony. No faction could impose its vision, however, and active collaboration with German demands made any planning an academic exercise. Still, a consensus existed that the state needed to forge a new political settlement and break with old patterns of authority.

The wartime Resistance movement also wanted to renew France. It shared with the avowed enemies at Vichy the criticism of Third Republic politics, the fear of liberalism, the need for state planning, and the hope to depoliticize agenda setting.[3] In contrast, however, the Resistance favored greater democratic participation and social responsibility. The diverse tendencies in the Resistance sought to reform economic institutions and to eradicate "Malthusianism"—the catchall phrase for weak entrepreneurship, inability to innovate, failure to take risks, and economic pessimism. Development of a coherent program was ultimately impeded by the tensions among enlightened business leaders, Christian Democrats, Socialists, and Communists. Wartime rhetoric forged the critical political compromises, but it could not substitute for coordination nor conceal the differences between the adherents of state-led and market-led solutions to economic renewal.

Neither Vichy nor the Resistance had an opportunity to enact vision of economic renewal. The Vichyites had no political credibility after the German occupation of southern France in 1943 and the total German defeat in 1945. The Resistance coalition was too divided to impose a clear program on the Liberation governments. A confluence of the dynamics of party politics and social forces enabled postwar leaders to carve out a set of institutions to fulfill that vision of industrial renewal. In spite of other partisan differences, various political forces agreed to forge a planning commission, isolate it from ministerial and parliamentary pressures, and staff it with sufficiently like-minded individuals committed to modern production methods and efficient allocation of resources. Organized labor and the left vigorously supported the process of "indicative" planning—establishing economic targets and using policy instruments to achieve them.

Behind the partisan politics of the Liberation and the early Fourth Republic were enlightened business leaders and high civil servants who argued that planning was the only substitute for nationalization. After the 1930s, the CGT, some Socialists, and the forces around Pierre Mendès-France had argued the complementarity of planning and nationalization: state control of key resources would facilitate state-directed growth. De Gaulle's dismissal of Mendès-France from the Finance Ministry in 1945 temporarily censured that option. A centrist group of modernizers around Jean Monnet then argued that while nationalization required displacement of property rights, planning implied a compromise—production and investment targets but also constant *negotiation* over methods of resource allocation.[4] To mobilize capital and skills, both business and labor needed to be included in the planning process.

This center group had considerable assets. The first three commissioners of the plan—Monnet, Etienne Hirsch, and Pierre Massé—had been respected business leaders. Business associations had been sufficiently discredited by their role during the war that they could offer little effective opposition. In addition, Monnet was an astute political player with an extraordinary international network who relied on strong personal convictions and was capable of effective negotiation.[5] He packaged together American ideas, bureaucratic resourcefulness, and an invigorated state apparatus.

Monnet had contacts with leading U.S. industrialists and political figures.[6] In the immediate postwar period, he used American funding to finance key projects and American know-how to organize them.[7] Afterward, the relationship between planners and the Americans facilitated the opening of the French economy. Both the Truman and Eisenhower administrations encouraged the development of the ECSC. The "American way"—that combination of entrepreneurial robustness, economies of scale, and organizational rationality—fascinated French modernizers.[8] The First Modernization Plan depended on Marshall Plan funding.

The postwar reforms increased the resources in the hands of state planners. Nationalization of coal, electricity, and railroads gave the state leverage over the energy and transport necessary for other economic activity. Although intended to penalize wartime collaboration, the takeover of Louis Renault's firm gave the state an important position in the market for automobiles—the product that would spur postwar manufacturing innovation. Nationalization of deposit banks and insurance companies brought control over credit allocation. The key to postwar intervention was the sophisticated financial system that facilitated the channeling of investment and active discretion in the use of funds.

Postwar economic leaders harnessed existing state machinery. The sectoral committees of the First Modernization Plan assembled state bureaucrats and industrial managers in ways that resembled the Organization Committees of Vichy. Planners acquired the financial tools of the Third Republic and the Vichy experiment. The Bank Act of 1945, which nationalized the Bank of France, continued the process of socializing the central bank that had begun with the Popular Front. Vichy had been responsible for shifting power from the major credit institutions to a banking committee, which would facilitate the postwar *encadrement de credit*—quantitative credit restrictions.[9]

The interventionist state already enjoyed administrative coherence. The elite higher education system (the *grandes écoles*) provided the economic ministries with a core of professional civil servants. Through the phenomenon of *pantouflage*, graduates moved into the business community after state service, carrying with them the informal networks of their *corps* and concrete knowledge of policy-making processes. The domination of this state-created elite ensured a degree of homogeneity at a cultural level seldom found in other advanced countries.[10]

The centralization of administrative authority gave the state "eyes and ears" in the provinces. Not only did the local prefect oversee municipal budgets and police forces, but also the office of the prefect served as liaison between the ministries in Paris and local industries. Departmental labor inspectors responsible to the Ministry of Labor monitored industrial relations and could intervene in disputes. The spending ministries (Industry, Housing, Public Works) had local representatives in the field responsible to Paris.

Centralization of power by itself did not bring the capacity to intervene. The planners used various strategies to strengthen their assets. First, they developed the political will to change society. They gained such resolve through a rhetoric of renewal sufficiently powerful to discredit potential losers—either the traditionalists or the anticapitalists. They benefited from the postwar mobilization that precluded a return to prewar economic orthodoxy. With the sectarianism of the cold war, the PCF and CGT dropped out of the adjustment model. The PCF was expelled from government in 1947, while the CGT claimed that planning favored only capital. Internal pressures within the Socialist Party (SFIO) and the Mouvement Républicain Populaire (MRP) precluded radical redistribution. Therefore, with the Left and the Right hamstrung, planners had freer hands to forge an interventionist model.

Second, openness to international markets became a wedge used by modernizers to encourage competitiveness at the level of the firm. The ECSC constituted the first step toward a European solution to financial solvency. Steel and coal were two of the six sectors targeted by the Monnet Plan and considered keys to industrial renewal. Ironically, Monnet envisioned the ECSC as a political solution to the Ruhr question and only secondarily as a spur to French industry. Yet once the process had begun, openness to markets pressured steel makers to modernize plants and equipment and to combine assets. It also spurred the rationalization of the coal mines.[11]

The ECSC gave the modernizers important political ammunition when it began operation in 1952. Monnet argued that the steel men, symbols of Third Republic Malthusianism, sought to maintain understandings among companies that carved up a protected national market.[12] (He played down cost advantages enjoyed by the Germans and Belgians.) Steel interests could not ally with coal to oppose the Treaty of Paris, since the Charbonnages de France belonged to the nationalized sector. Instead, it found "objective allies" in a PCF and a CGT who complained about a new German "invasion" and a "Europe of trusts." This ad hoc national coalition failed to defeat the treaty. It was further weakened by the Treaty of Rome, which instituted the European Economic Community in 1958.

Third, modernizers worked to change the bureaucratic balance of power within the state to avert such a blocking coalition. The state-led adjustment model pitted the traditionalist Ministry of Industry (MoI) against the modernizing Ministry of Finance (MoF). The MoI traditionally served as *tutelle*—guardian—for various economic sectors, combining active promotion of the industries it supervised with a technical understanding of market directions. Consequently, the MoI had close ties with industry and was perceived as an industry advocate, unable to execute the types of discriminatory treatment necessary for the interventionist state.[13] Monnet transformed the prewar conservative MoF into the dynamic actor of the postwar period.[14] This effort needed the reinforcement of the Trésor (Treasury) and the construction of the interventionist apparatus, all motivated by a small elite dedicated to industrial reorganization. The latter had been converted from the orthodoxy of the 1930s and was staffed by individuals, such as François Bloch-Lainé, who were cognizant of the Keynesian revolution that was transforming modern economics. The policy "player" became the MoF which, in conjunction with the Planning Commission, defined the path of strategic intervention in the early period.[15]

In spite of bureaucratic rivalry, the concentration of policy tools facilitated the redirection of priorities. Bureaucrats and industrial clients resisted the decline of the MoI. Eventually business interests sought influence in the MoF, thereby weakening the interference of the MoI. Later conflicts took place between the MoF and the Planning Commission over the implementation of plans. Each actor in the system—from nationalized industries and banks to the MoF—has its own priorities. Yet the French state developed the capacity to think strategically and act coherently, although its policies have sometimes failed in both respects.[16]

Modernization proved a conflictual process. From the beginning it was combined with the tasks of postwar reconstruction to encourage increased production, new production technologies, and new patterns of authority relations. Delegates who made missions to the United States were astounded to discover human relations management.[17] According to Philippe Mioche, Monnet sought the unions' participation in the planning process.[18] It was not easy to encourage organizational innovation. French managers found it difficult to accept less than complete subordination of labor at the workplace. French management circles were too divided between small and large companies and between family and corporate structures to speak with one voice.

During the Fourth Republic (1946–1958), careful development of a civil service elite provided relative stability in economic management. This had been a key goal of Monnet and his team from the beginning: both the Planning Commission and the Trésor were deliberately kept small to avoid bureaucratic infighting. Bureaucratic stability was particularly important with the political turmoil of the 1950s. The struggles over the colonies and France's position in the world were poignant testimony to the absence of political consensus on key issues of the day.

Stable rule, however, did not mean that planning was apolitical. As Stephen Cohen remarks, "Planning is either political or it is decorative."[19] In the Fourth Republic, planning involved the use of different forms of coercion to implement state priorities (to be discussed later). Likewise, ministerial offices were staffed by a civil service elite who thereby received an initiation into politics. Moreover, as Ezra Suleiman notes, because French law allows bureaucrats to serve in elected positions while maintaining their status in the civil service, the distinction between politicians and bureaucrats was blurred.[20]

With the concentration of state power in executive hands in the Fifth Republic in 1958, party politics always challenged state-led growth. When planning priorities changed in the 1960s—away from basic sectors, heavy industry, and import substitution to more social-oriented policy—economic planning could not remain divorced from the electoral concerns of French politicians. Prestige projects were introduced into the planning process by Gaullist governments with clientelistic implications. Exclusion of the unions brought active disdain from the Left before 1981, and consequently the process was delegitimized. The polarization introduced by the Fifth Republic politicized every act of industrial industrial policy making. If leaders could take credit for success, they could also be blamed for failure.

Likewise, concessions to traditional sectors provided the electoral support from the very groups—shopkeepers and peasants—who were threatened by state-led modernization. A frontal assault on *la France profonde* would have been suicidal for the modernizers. It was difficult enough to encourage capitalists to invest and shift resources to new industries. The laws restricting commercial development in favor of small shop owners bought the political space to pursue industrial policy.[21] Subsidies to French peasants—both from Paris and via Brussels and the European Community's Common Agricultural Policy—continued to be means whereby resources were transferred from one economic sector to another.

Finally, policy tools could be used for goals that contradicted modernization. The interventionist apparatus could prop up "lame ducks" as well as encourage competitiveness. Industrial policy in the 1970s consisted largely of subsidizing large and small firms that suddenly found themselves in competitive difficulties.[22] Labor market policy had to cope with the pressures for redundancies, leading to increasingly costly payments to workers for not working.

After 1945, the French state embraced centralization as a means to transform society. A new generation of leaders filled key positions in the bureaucracy. They built new and fine-tuned old institutions to pursue their agenda. State-led adjustment, however, was never above reproach. It enjoyed enormous cultural capital and had a powerful institutional impetus, yet it remained vulnerable to political unpredictability.

State-led adjustment was all the more precarious because it was incomplete. The organization of indicative planning required constant negotiation among economic actors. However, concessions to traditional forces eventually created a mismatch of policy tools: the state developed policies for mobilizing capital only by relinquishing control over labor policy to the business community. The effects of company-led labor policy on productivity were not felt in the early period after the war because production was low-skill. By the 1960s, however, labor's exclusion and working-class mobilization encouraged protest against the state-led model.

Policy Tools for Promoting Industrial Growth

Modernizers developed impressive tools with which to shape industrial policy after 1945. Control over credit allocation enabled the state to influence investment choices. Subsidies in various guises both induced cooperation and built a political constituency. The Planning Commission generated consensus among the ministries and within the business community over the direction of industrial change. Both financial control and consensus building facilitated the transfer of resources from traditional to growth sectors while maintaining relative political stability. The constitution of the Fifth Republic provided additional political equilibrium by centralizing power in the executive.

The MoF was at the apex of industrial policy making. It directed both monetary and fiscal policies at the macroeconomic level. Its importance for industrial adjustment rested with the scope of its responsibilities. Although ministerial organization has changed several times after 1945, the MoF consistently combined bank regulation, control of equity markets, price controls, and supervision of nationalized industry. It also ran the Economic and Social Development Fund (FDES), a major source of low-interest investment funds. The powers of the MoF grew so large that President Valéry Giscard d'Estaing halved it in 1978 and created a separate ministry for state budgeting functions.

Within the MoF, the Trésor has coordinated macroeconomic policy and set the agenda for industrial policy.[23] With a small staff, the Trésor coordinated the work of the MoF and served as its "nerve center." Its power derived from a capacity to persuade other departments inside and outside the MoF to engage in particular forms of behavior and to "mobilize bias." Perhaps its most powerful tool for industrial intervention was the FDES. The fund has been run by an interministerial committee whose subcommittees are chaired and controlled by Trésor officials. The allocation of low-interest loans induced firms to undertake projects in line with state planning.

The Trésor managed the mechanisms of selective credit allocation. For a variety of historical reasons, French securities markets were comparatively weak after 1945. Consequently, firms turned instead to the banking system for investment funds. The Trésor has been deeply implicated in the operations of that financial system. A number of parapublic financial intermediaries have been tied to the MoF. For the most part, they borrowed funds from the commercial banking system and channelled them to industry along the lines approved by the Trésor. The Crédit National directly loaned and guaranteed funds to banks who in turn extended medium-term loans to industry. It also managed the funds extended by the FDES. The Caisse des Dépôts et Consignations used funds from the public savings network to buy loans and securities. The Crédit Agricole drew savings from rural cooperatives and invested them partially outside the agricultural sector. The Trésor either directly controlled or influenced these three financial institutions, defining policy agendas and supervising the course of investments.

The Trésor also controlled activity in the small securities markets. In the bond market, it set the interest rate on public issues, which historically have dominated activity. Prices were kept artificially low, thereby encouraging an overdemand for the available supply. This forced firms to "queue" for the right to issue bonds, again giving the Trésor enormous discretion. Nonmarket criteria influenced the stock market as well. New issues needed administrative approval in the primary market, and the banks played a major role in the secondary market.

The commercial and investment banks also had close ties to the MoF and the Trésor. The major commercial banks—Crédit Lyonnais, Société Générale, and BNP—were nationalized by the postwar government. The two key investment banks (Paribas and Banque de Suez) remained private, although they were nationalized by the Socialists in 1982 and privatized by the Conservatives in 1987.[24] Nonetheless, both the commercial and investment banks borrowed funds from the financial intermediaries, giving the latter (and ultimately the MoF) leverage in the final allocation of credit.

These financial tools were used to shape industrial policy after 1945. Over the long term, modernizers sought to reform risk-avoiding capitalists. On an individual basis, such "backward" behavior was not extraordinary, given that conditions for capital accumulation were not always propitious.[25] What distinguished French capitalism was that political coalitions reinforced such attitudes and impeded the mobilization of capital.[26]

In the shorter term, policy tools could be used to influence, prod, and cajole industry to invest in particular directions and adopt new production technologies. The two-step process from savings to investment—from deposit to intermediary institutions, from the intermediaries to credit institutions—enabled the MoF to play a large role in mobilizing finance. It could choose who would receive financing, especially at favored rates. The nationalization of the major insurance companies in the immediate postwar period brought further control over capital markets. With such muscle in the market for investment funds, the Trésor could "make policy that is intended to shift resources between sectors and even between firms."[27]

Because it was small, however, the Trésor needed administrative allies to coordinate strategic action. Within the MoF, the Direction Générale des Prix administered price controls on thousands of products until the mid-1980s. It targeted mass consumption and investment goods in efforts to control inflation. Through its policies of exceptions from price controls, the MoF influenced the behavior of individual firms. By the 1960s, exceptions had been granted by the thousands.[28]

The policy of exceptions extended to fiscal policy. All governments use tax codes to spur investments while granting exemptions on political and economic grounds. The French system, however, used ad hoc procedures more often than most. The MoF collected taxes and allocated relief. It granted relief from corporate income, payroll, and value-added taxes in exchange for quid pro quos from the private sector. Those actions might include higher exports, investment in targeted projects, or job creation. Tax relief, Shonfield notes, was "part of a larger complex of financial arrangements that help to ensure that French investment keeps moving in the direction and in the volume required."[29]

Like the Trésor, the Planning Commission had a small staff, and much of the work took place in committees staffed by outsiders. These brought together representatives from business, labor, and the state for particular economic sectors (in "vertical" committees) or for broad-based policy issues, such as employment and research (in "horizontal" committees). Labor participation was insignificant, for the unions were either marginalized from key decision making or they withdrew in protest from the process.

According to Peter Hall, the planning process served three major functions. First, it built alliances between the business community and state actors. Second, it laid out the economic agenda facing the nation by narrowing the range of possible choices. Third, because the agenda sought to transform France into an international competitor, the process had to socialize participants into a belief that change was possible *and* in their interest.[30]

Planning not only politically rehabilitated the business community in the early postwar period, but also provided the corporate community with tools for economic expansion. The planning committees encouraged the exchange of information between corporate managers and state bureaucrats that facilitated demand forecasting. It provided a permanent dialogue that helped allay mutual suspicions.

The Planning Commission, especially in the early period, worked closely with the MoF to set the terms for adjustment and to channel resources to favored areas. The Trésor had the necessary resources to implement the Plan.[31] This relationship implied a division of labor. The MoF engaged in both long- and short-term policy making, while the Planning Commission focused on long-term goals. It was also more insulated from political pressures than the MoF, although the Trésor itself enjoyed a similar respite.

Setting priorities meant adapting to contextual exigencies. Shortly after the war, the First Modernization Plan sought to remove bottlenecks in basic industries by targeting investment funds to six key sectors—coal, steel, transportation, electricity, cement, and agricultural machinery. After reconstruction, the Planning Commission endeavored to generate growth in the bulk of manufacturing (1953–1961). Later it promoted mergers, international competitiveness, social infrastructure, and applied research. The Plan played a relatively minor role in industrial policy after the mid-1960s, and the MoF assumed control of industrial policy making. By that time, however, its major work had been accomplished: French industry—at least the favored firms—had been shoved into the modern era.

Business inclusion in the planning process shaped the character of industry's policy choices. In the early stages, business resisted labor participation. This was particularly true after French economic growth took off in the 1950s. Later the size of large companies altered the capacity of state intervention. The policy of national champions—consolidating industrial assets into a small number of large national companies—created firms that were less amenable to the policy tools of the MoF. During the period of high unemployment after 1973, governments set up a number of agencies to subsidize companies on the verge of bankruptcy. In the 1980s, industrial policy focused on developing technological policy cores and creating strong and autonomous financial institutions to work closely with industry.[32] Perhaps not surprisingly, the cultural elite from the *grandes écoles* remained entrenched in both the state and business.[33]

The French state has long shown the capacity to influence behavior, to influence the policy agenda, and to generate support for its policies. Through control of the financial sector, the MoF targeted investments to growth sectors in the economy. The Plan prepared French industry for the rigors of international competition introduced by the Common Market. Yet the state was not strong enough to impose its will without concern for the consequences. Subsidies to agriculture and small business forestalled conflict over the adjustment model. Inflation was used as a conscious policy tool to reduce the real value of those subsidies, thereby masking the intentions of policy makers.[34] The Gaullist coalition between 1958 and 1974 trapped politically many of those groups most damaged by the new political economy. In the process, it developed obligations and involvements that reduced its own freedom of action.

The Slow Development of Active Labor Policy

The priorities of the reconstruction period favored capital. While business was included in planning, labor was marginalized. The perpetual isolation of the unions from policy making transformed short-term economic necessity into longer-term political domination. State leaders who tried to influence the price, allocation, and quality of labor saw their influence and policy levers develop very slowly. As a result, the economic development process envisioned by the modernizers was slanted away from broad cross-class inclusion.

The initial intent of the Monnet group was to include labor to some degree. Product market Malthusianism, it was argued, both derived from and encouraged labor market paternalism—traditional labor-management relations that depended upon reciprocity and loyalty between owners and workers rather than market signals. Therefore, both needed to be tackled to spur growth. As suggested above, Monnet envisioned dealing with organized labor on a regular basis. François Bloch-Lainé—head of the Trésor between 1947 and 1953 and afterward director of the Caisse des Dépôts—sought to "reform the firm" by proposing an economic citizenship through giving unions a participatory role at the workplace.[35] Still, for bureaucrats and ministers, enhanced production took precedence over new forms of labor inclusion by the early 1950s. Transforming French capital in both product and labor markets would have been a Herculean task that even Monnet's team could not achieve.

Concessions to traditional forces, critical to sugaring the pill of industrial modernization, further retarded the state's capacity to intervene in labor markets. The party coalitions that shifted within narrow bounds in the Fourth Republic further slowed the state's actions on labor policy. With Communists and Gaullists as pariah political forces while drawing 40 percent of the vote, the remaining parties could not alienate local business leaders. Therefore, they struck a bargain: in return for cooperation with investment targets, employers could have relative freedom in labor relations.

Given these conflicting incentives, the critical question was under what conditions and with what tools the state would intervene in labor matters. Was intervention in industrial relations needed to quell social unrest, a means of economic redistribution, or a corollary to a competitiveness strategy? Each goal had its own adherents, corresponding roughly to the three groups of modernizers—conservative, left-wing, and centrist.

The state played a mixed role in labor policy until the 1980s—entering some arenas, avoiding others. It set the legal framework for the employment contract and intervened in labor-management relations, but it left employment, incomes, and skill policies to company discretion. Labor market measures were "passive," encouraging market entry or exit. Consequently, state policy was more reactive to social disturbances than proactive in shaping the labor market. It conformed to, rather than shaped, market behavior.

A complicated system of labor law developed in France. Given the employers' disdain for unions, state regulations proliferated to protect the employment contract. Because of weak unions and employer associations, collective bargaining focused more on industrywide minimum wage rates than on other issues. Therefore, the Ministry of Labor played a crucial role in determining the framework of the employment contract, provisions for hiring and firing, working hours, and protection against discrimination.

The Ministry of Labor played a key role in collective bargaining. Legislative frameworks have determined who can bargain and over what issues. Legislation and government decrees on various social policies—paid vacations, sick pay, working time, retirement pay—both delimit and influence the content of collective bargaining. The ministry can extend privately negotiated agreements to an entire region, to an industry, or to all wage earners. It can also force social partners to negotiate by threatening state action if bargaining does not take place or succeed. The departmental labor inspector has wide-ranging powers to intervene in local industrial conflicts. "If there is a 'web of rules' about industrial relations in France," writes George Ross, "then it has been predominantly spun by the state."[36]

In contrast, the French state has been much less successful in developing incomes, employment, and training policies. Attempts to refine labor market instruments did not come until the 1960s, although the state did influence wage negotiations through the use of a minimum wage tied to both the cost of living and macro productivity increases. This succeeded partially because economic growth in the 1950s increased employment and earnings. Also, the types of jobs to which individuals migrated did not have high skill requirements.

In the early 1960s, the Gaullists attempted an incomes policy by controlling public-sector wages so as to set an example for the private sector. The strategy broke down in the face of unified hostility from the unions. Labor had little reason to cooperate since it was too divided to make agreements stick and the state was unwilling to confront traditional attitudes in the private sector.[37] Therefore, until the mid-1960s, the state only monitored wage levels. Afterward, it forced companies to divulge aggregate salary levels before permitting price increases. Incomes policies gained effectiveness only in the 1980s through stricter public sector guidelines and union demobilization, and this largely due to high levels of unemployment.

The tools of employment and training policies were forged in the mid-1960s as a corollary of the push for national champions. The Agence Nationale pour l'Emploi (ANPE) was founded in 1967 to provide centralized job placement. (Previously, local labor markets were served by local manpower bureaus.)[38] The ANPE had four major tasks: to match available workers with available jobs, to direct certain categories of unemployed workers toward job training, to orient eligible job searchers toward public benefit programs, and to garner statistical material. The agency had to work closely with employers to determine the scope of local labor markets. This meant interviewing and counseling job seekers to determine their skill levels and employability.

A nationwide system of unemployment compensation was not created until 1958 and did not provide near universal coverage until 1967.[39] The Union Nationale pour l'Emploi dans l'Industrie et le Commerce, (UNEDIC) receives funds from the payroll tax. Through local affiliates, the Associations pour l'Emploi dans l'Industrie et le Commerce (ASSEDICs), the UNEDIC pays benefits to insured workers with deficits covered by subsidies from the Ministry of Labor. Employers contribute relatively heavily to the insurance fund, paying two-thirds against employees' one-third.[40] Since the UNEDIC system derives from a collective bargaining agreement, and is essentially a private union-management insurance agency, coverage and orientation change with new negotiations. The state can also encourage unions and employers to revise contributions, benefits, or other aspects of the system. The UNEDIC coordinates its activity with the Ministry of Labor and the ANPE. The state has had considerable leverage over the terms of insurance—coverage and duration—since it finances the deficits out of general tax revenues.

The French training system is relatively dispersed. The Association pour la Formation Professionnelle des Adultes (AFPA) was created in 1966 to run roughly 130 training centers and subsidize others. The backbone of the training system consists of an employer payroll tax whose proceeds go toward adult training. This system resulted from a peak-level agreement between the major unions and the peak employers' association—Confédération Nationale du Patronat Français (CNPF)—in 1970 and the law that codified it in the following year. Although typically referred to as the "1 percent tax," the precise rate varied after 1971, constituting 1.2 percent of the salary mass by the late 1980s. The tax is paid into a training fund (the Fonds d'Assurance Formation) run by the training division within the Ministry of Labor—the Délégation à la Formation. The FAF, in turn, pays for training and remuneration of participants, although firms deduct training expenses from their contributions. The training takes place in different types of private, public, parapublic, and nonprofit organizations as well as internally within the firm itself.[41] This financing system enables large firms to pay little on average into this fund, since they are most likely to run their own in-plant training. Instead, it socializes costs among small and medium-sized producers.

The key to worker adjustment has been the Fonds National de l'Emploi (FNE), a flexible interministerial fund started in 1963.[42] It evolved from being a facilitator of labor market equilibrium in the 1960s to a critical instrument of industrial restructuring. Because it has never been clearly earmarked for specific programs, it can be used for a wide variety of purposes—early retirements, training programs, supplemental or partial unemployment benefits, or reductions in working time. Policy makers can use the FNE either as seed money in conjunction with other funds or as the single financier of a given program. The strength of the FNE comes from its capacity to serve as a policy lever well beyond the financial impact of the funds involved.

By the late 1970s, this leverage was used in negotiations for public financing of firm-level employment reductions. A firm or industry typically signed an agreement with the FNE through the Direction Départementale de l'Emploi (DDE).[43] The terms of the FNE's financing of certain provisions—for example, early retirement or training programs—might include requirements to offer younger workers new jobs within the company to keep the numbers of early retirees down, to spread out employment, and preserve the age pyramid. The Délégation à l'Emploi (DE) within the Ministry of Labor uses the FNE on a case-by-case basis depending on the solvency of the firm, the behavior of social actors, and the reabsorption capacity of local labor markets. As the wave of redundancies hitting French industry became politically embarrassing in the mid-1980s, the DE rethought the manner in which it could leverage FNE funds and sought different ways to keep the numbers off the unemployment rolls. The Ministry of Labor can also sanction noncompliance by reducing post facto the FNE's share of financing, thereby increasing the firm's costs.

The Ministry of Labor has also used the FNE in conjunction with placement, training, and unemployment compensation agencies. As a co-financer, the FNE introduces flexibility to programs that otherwise might fall prey to the rigidity of established bureaucracies. It can overcome the budgetary constraints of individual agencies. In its agreements with other state organs, however, the FNE has trouble exercising the leverage it uses with private industry or even the nationalized sector. It has even been criticized for acquiescing too readily to poorly designed employment programs.[44] This interministerial fund is, therefore, subject to interministerial jockeying.

While the French state had developed a set of discretionary tools for industrial policy, it had blunter instruments for labor market policy—in spite of the use of FNE funds. As long as workers left the land for semiskilled employment, this imbalance between sharp product market and dull labor market tools did not create a bottleneck. The development of national champions led to a coordinated national labor market policy, but the institutions created—the ANPE, UNEDIC, AFPA, and FNE—required little active intervention in either labor management at the firm level or proactive employment policy. Blunt policy tools created problems if the state wanted to intervene consistently in labor markets, as it did by the 1970s, to alter the quality of labor skills and to spur growth in new economic activity.

The political tenacity of small business weakened the coordination of labor policy.[45] Powerful mechanisms for channeling funds to industry and blunt labor market instruments subverted the modernizers' ambitions to allocate labor resources more efficiently. The absence of regular venues for including organized labor in negotiations preserved those traditional forces that modernizers had hoped either to convert or replace. The failure to incorporate labor both skewed outcomes and encouraged social protest.

Labor Market Organization

The state's capacity to forge a more interventionist labor market policy was also stymied by the labor movement. On the one hand, a deep hostility to capitalism has long permeated French trade unions. On the other, French working-class representation has been refracted by competition; workers can choose from an array of organizational options to express their political perspectives or articulate plant-based grievances. The exclusion of labor by business was reinforced by the class-based nature of the French labor movement and the incoherence of labor organization. Even when the state or employers have sought to include labor, the choice of what union to negotiate with has always been a dilemma.

The divisions within French labor are deep. The CGT was hegemonic in most industrial sectors until the 1960s, although its history had been sprinkled with division and secession. Born of the anarcho-syndicalism of the nineteenth century, the CGT disdained political affiliations, codifying that disdain in the Charter of Amiens in 1905. As with socialism in France, the French labor movement split over the conditions of the Third International. A revolutionary faction formed the Conféderation Générale du Travail Unifié (CGTU) in 1921, which worked closely with the PCF. The Popular Front reunified the CGT, with the moderates maintaining control. Only the war, the Resistance, and Liberation dislodged the moderate forces in the confederation. At the 1946 congress, the former CGTU faction gained organizational control.

The PCF and CGT have had close relations since 1946. At the national level, the secretary general of the union typically sat on the political bureau of the party, and several union officials have party responsibilities. Perhaps 10 percent of union members belong to the Communist Party.[46] In spite of Communist leadership, the confederal council has always had a minority of Socialists. Both at national and local levels, however, official policy rarely diverged between union and party. The "transmission belt" tended in the 1950s to run from the PCF to the CGT: the party used its leadership positions within the union to make labor market actions complement and conform to Communist Party policy, although party dominance lessened after the early 1960s.[47] Afterward, the transmission belt could not function in a complex social situation that demanded that unions remain independent. Instead, the union gained "relative autonomy" from the party. Nonetheless, the CGT remained ensconced in class-based action—a *syndicalisme de masse et de classe*.

The PCF contributed to the analytical positions of the CGT. Given the consonance of world views between the dominant faction of the CGT and the PCF, the two organizations could borrow ideological positions and labor analyses from one another. If one compares the publications of the party and the union press from the 1950s, one finds similar perspectives on the "pauperization" of the working class, the dangers of the Marshall Plan for French sovereignty, the threat of U.S. imperialism, and militant support for the peace movement and friendship with the Soviet Union.[48]

The FO broke from the CGT in 1948 in opposition to the close relations between the PCF and the majority faction of the CGT. (Likewise, teachers broke from the CGT in the same year.) While the CGT favored state action, the FO conceived of union behavior in marketplace terms. The new confederation grouped together trotskyites, anarchists, gaullists, socialists, and even a few "independent" communists.[49] The only unifying political sentiments that united the FO were anticommunism and "bread-and-butter" demands.

Catholic unionism has a long history in France. Founded in 1919, the Confédération Française des Travailleurs Chrétiens (CFTC) offered moral solidarity to Catholic workers. It split in 1964 over political tactics and organizational forms. A rump of Catholics maintained the old organization of the CFTC but were concentrated in isolated economic sectors, such as coal mining. After the 1964 split, the union was more oriented to an apolitical localism. Of the four industrial blue-collar unions, it has had the smallest membership and fewest votes in workplace elections.

The CFDT emerged from the CFTC as the larger organization. Strongly influenced by opposition to the Algerian war and the activity of the new Left, it sought a unionism different from the "mass and class" strategy of the CGT. The CFDT has had a mixture of identities: Christian social, decentralization—it was the first union to adopt the rhetoric of *autogestion* (self-management)—employment solidarity, and collective bargaining.

The Confédération Générale des Cadres (CGC) was formed in 1945 to represent the interests of midlevel managers, engineers, and technicians. Formed around a core engineering union that had not been dissolved by the Vichy government and relying on a tradition of friendship organizations, the CGC saw itself as a middle-class group squeezed between industrial unions and upper management. It signed collective bargaining agreements for its purview, which sometimes overlapped with the other unions. Its party contacts have been stronger on the right than on the left, and it has sought to differentiate itself from the blue-collar unions.

Relations among these five unions vary from tense to hostile. Politics matters considerably in interunion relations at all levels, and political divisions do not run along abstract ideological lines. Surveys indicate that militants tend to be party members and very interested in national politics.[50] Each political orientation carries with it a world view and an arsenal of tactics to achieve certain goals.

The PCF-CGT relation has not been the only party-union connection in France. The FO defines itself as vaguely socialist (in spite of its heterodox composition), and until the Communist-Socialist alliance at the national level in 1972 had cordial relations with the predecessor of the Socialist Party, the SFIO. In the late 1940s, it benefited from considerable state subsidies from SFIO ministers.[51] Relations between the CFTC and the remnants of French Christian Democracy (the MRP) were also close, at least immediately after the war.[52] The CFDT found itself close to the Parti Socialiste Unifié until the PSU merged with the Socialist Party (PS) in 1974. The connection between the CFDT and the PSU faction of the PS has led certain commentators to speak of a "second" Left, one that is less statist, more decentralizing than the "old" Left.[53]

The PCF-CGT nexus differed from others in the organizational capacity of the PCF and its role as a political outsider. Until the 1970s, the PCF was the largest political force in France at both the national and local levels, while the CGT remained the largest union confederation. The confluence of sheer organizational resources enabled the union and party to maintain a large mobilizational effort. The tight linkage between union and party gave the PCF-CGT relation an important grounding in working-class culture. As an oppositional political force that was excluded from governing between 1947 and 1981, the PCF provided strident critiques of French capitalism and state-led modernization. Its articulation of working-class positions helped sharpen the expression of class conflict among French workers.

Political cleavages complicate joint action but do not preclude it. Local militants have their own histories. Traditions of respect or animosity develop according to individual patterns of interaction. National alliances facilitate relations at lower levels. Strike committees bring together militants from different unions for limited and focused combat. The more common norm, even in periods of confederal cooperation, has been quiet aggression.

French unions compete with each other to represent a small percentage of the work force. Except for surges surrounding the Popular Front and Liberation, nationwide membership averaged between 20 and 25 percent of nonagricultural workers in the 1960s and 1970s—split among the five industrial unions, the teachers' union, and independent unions. In the late 1980s, it sank below 10 percent. Some industries have few union members, for several reasons. First, until the mid-1960s, employers tended to view their management prerogatives as a divine right. Second, until 1981, the French state clearly sided—if only by failing to enforce key aspects of French labor law—with employers in excluding labor from the shop floor. Third, continual electioneering on the shop floor creates a free-rider problem: employees can vote for a union and be covered by union-negotiated agreements without paying dues.

Compounding the problems of low membership and pluralism is union decentralization. Each confederal union has several decision-making points. At the top of each union movement is the confederation that coordinates interfederal relations and engages in political-level negotiations. Each confederation generally organizes federations along industry lines—for metals, chemicals, docks, etcetera. Federations within the same confederation do not compete within given industries for the workers' loyalty.

Both the confederation and the federation have local-level organizations. The confederation organizes *unions départementales* (UDs) at the departmental level and *unions locales* (ULs) at the town level, or at the neighborhood level in larger cities. The UDs represent all union members, regardless of federation. The ULs represent workers across sectors, but tend to be staffed by militants from a dominant firm. These ULs report to both the UD and the federation. After the Grenelle Accords of December 1968, the federation organized plant-level sections (*sections syndicales d'entreprise*, or SSEs) that could deal directly with workers at the workplace. Regional bodies frequently coordinate actions among different SSEs and negotiate with employers.

The method of dues collection in French unions accounts for both their poverty and confusion over the actual number of unionized workers. Union militants from the SSE collect monthly dues. No check-off system exists in France, although unions began to use automatic bank withdrawals by the 1980s. A union member typically buys an annual card with space for twelve stamps, and union militants sell stamps on a monthly basis. Although large variations exist among the confederations, workers buy an average of under twelve stamps—ranging from roughly ten for the FO to six for the CGT. So who is a union member—one who buys three, six, nine stamps, or one who simply buys a card?[54] Union figures are notoriously unreliable.

Unions compensate for organizational poverty with revenues from external sources. They receive state subsidies for training militants, and their press receives funds targeted for the press in general. Militants who sit on regional councils, the Economic and Social Council, and other appointive bodies return salaries to the union. Since union members tend to sit on works councils, there is a temptation to use those funds to promote union ends.

Nonetheless, French unions are too poor to amass significant strike funds, and this has had organizational consequences. This inability to sustain workers during strikes further decentralizes union action because the federation does not have the internal funds to regulate behavior. The monthly sale of membership stamps provides constant contact between the militant and members, who tend to have greater local than national loyalties. The lack of a strike fund also forces the union to maintain important community relations. The community of interests between the Communist Party and the CGT has helped to mobilize resources. These aspects of union organization facilitate the rapid translation of grievance into protest.

While membership figures are a closely guarded secret, the balance of union forces at the plant level is constantly tested. Annual elections decide the workers who will be *délégués du personnel*. Unlike U.S. or British shop stewards, *délégués* present the grievances of all workers in a particular shop to managers. Biennial elections are held for the works councils. Although not as developed an institution as in Germany, the French works council holds some responsibilities and constitutes a limited channel of communication between the work force and management. Because *délégués* and works council members cannot legally be fired, the institutions protect unions at the shop level.[55] Until the 1970s, *délégués* and works councils constituted prime examples of symbolic participation—they could meet with the employer but had no levers with which to influence company labor policy.[56]

Other elections test the unions' broader popular appeal. Unions present lists of employee representatives to the French social security system that disburses funds and collects premiums claims through tripartite boards of employers, employees, and state officials.[57] Every five years unions present lists for employee representatives in elections for bipartite arbitrators councils (*conseils des prud'hommes*) which handle worker and employer complaints against the other. Since 1984, one-third of the members of boards of directors (*conseils d'administration*) of nationalized firms have been workers elected through companywide elections.

Thus, several major electoral contests regularly test the unions' strength. Until the mid-1980s, the six confederations (five industrial and one educational) received votes from over 85 percent of voters in professional elections.[58] This fact casts a different light on the organizational weaknesses of the French labor movement. The French labor movement "represents" a higher percentage of workers than do movements elsewhere, although individual confederal unions do not have as many members. The French union almost assumes the attributes of a political party—with a core of militants, somewhat more members, and a vast constituency. At the local level, the union's mission is to articulate interests and get its representatives elected.

Elections also serve a legal function in French industrial relations. Any union that receives an average of 5 percent of workplace votes can be legally certified by the Ministry of Labor to negotiate in the industry. This certification of "representativeness" has been controversial. The CGT was deemed unrepresentative of railroad workers between the early 1950s and the late 1960s. When it gained representative status, it became the largest vote-getter in professional elections. The CGT fought the Ministry of Labor for years over the representative status of the FO in the steel industry.[59]

Because of the rules of representation, management could manipulate the process of collective bargaining. Until 1982 an individual employer or employer federation could sign a contract with any representative union. The union might receive only the 5 percent threshold, but the contract applied to whatever bargaining unit—plant, company, region, industry—the bargaining partners could decide upon. Management got a contract that least damaged its interests when a divided labor movement failed to present a common front.

Yet that manipulation does not lead to stable outcomes. Since contracts are not legally binding in French industrial relations, even the union signatory can strike after signing. The negotiating union also does not bargain in isolation.[60] It relies on the behavior of the other unions. The possibility of withholding a signature while pointing to the demands of others strengthens the bargaining position of the negotiating union. Employers are thus constrained by the knowledge that the work force may disagree with the terms of a contract. Consequently, the original negotiating union may be discredited in the process, in which case the employers would be forced to reach an agreement with a more aggressive union.

French industrial relations therefore contain important sources of instability. If the contract is not sacred and the bargaining partners fluctuate, conflict can result from other sources. Michelle Durand argues that mobilization can be viewed as a goal in itself, not simply an instrument of collective bargaining.[61] In a system lacking institutional consensus over the essential rules of engagement, conflict exists apart from contract demands. French workers and their representatives strike over negotiable and nonnegotiable issues.

A class-based, oppositional unionism survived intact into the postwar period.[62] This unionism saw the strike as a test of wills in the struggle for social change. Oppositional tactics posed a dilemma for French trade unions. They were consonant with world views of class exploitation and exclusion. Yet because they had the potential to assume a logic of their own, they risked the organizational integrity of the individual union. Consequently, French unions tried to channel such behavior with some apparent success—"skilled surfboard riders," as one group of authors term it.[63]

Likewise, the absence of regularized channels of bargaining with management encouraged unions to place their faith in political agents. Ideological congruency, therefore, was complicated by practical exigency. Incapable of promoting their interests effectively in the labor market, unions relied on their counterparts in party politics, further politicizing the process of state-led adjustment.[64]

Thinly organized French unions were forced to look to government for desired changes in production and working conditions. Lacking regular influence over government leaders, unions have pursued mobilizational strategies to induce state intervention. At the local level, unions have frequently relied on local leftist governments to subsidize meeting halls and soup kitchens, and even to provide some nonmonetary strike benefits.

Competitive unionism sustained mobilization until the 1980s. By contesting elections on a regular basis, union militants have tackled issues they might otherwise have neglected. This was particularly true of the FO, which, at the national level at least, strongly disdained strike actions. Consequently, the shop floor saw constant political agitation. Highly motivated militants used oppositionalist language to increase mobilization. Workers needed to be convinced that their interests were at stake and that they should engage in conflict.

Labor mobilization evolved throughout the postwar period. Variation has taken place along two axes, which we might label *mobilizational* and *organizational*. Figure 2.1 depicts the four possibilities that result from the intersection of the two. Mobilization has been political when programmatic demands relate only remotely to labor market action. These political demands reflect a confusion of responsibility between trade union and political party. Examples of political mobilization include "days of action" in which the unions organize walkouts on behalf of directives from the federation or confederation.[65] Political mobilization contrasts with labor market mobilization when unions focus on the conditions or terms of employment.

Mobilization also varies by organizational impetus. Who controls it—the confederation, the federation, or the local? The French labor movement may have decentralized decision making, but the center still seeks to exert control. Different forms of mobilization produce organizational stress. Locals chafe at directives from Paris, and the federation seeks to direct locally initiated mobilization. These differences are ordinarily masked by rapid organizational consensus building. Still, the efficacy of particular actions may not always be equally appreciated, a problem aggravated by competing strategies of different unions.

French mobilization has moved sporadically from cell I clockwise to cell IV in figure 2.1, below. Variation from political to labor market mobilization has depended partially on the willingness of labor organization to articulate grievances in concrete form. Despite the "oppositional" strategies of the CGT and CFTC/CFDT, both unions have

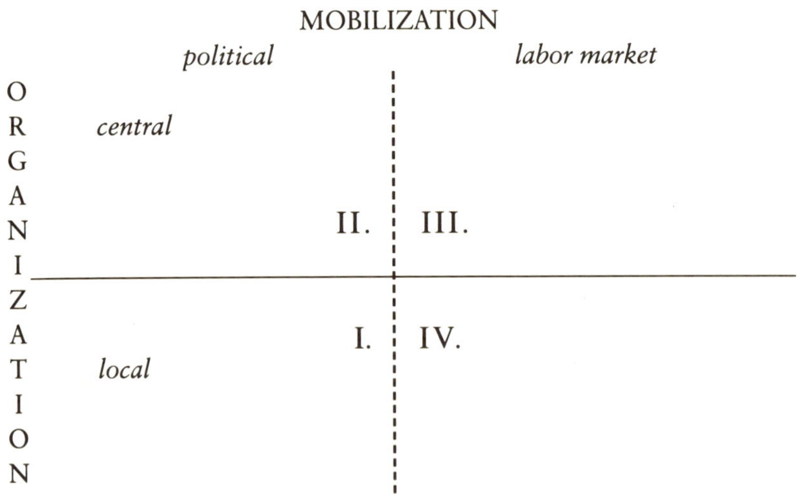

Figure 2.1 Forms of Labor Mobilization

sought at various times to dampen demands when they threatened electoral ambitions for the Left, negotiations with the industry, or state policy. This was evident while the Communists participated in the postwar governments between 1944 and 1947. The belief that a Left government was imminent in the 1970s also brought centralized mobilization.

The organization of labor interests in postwar France had at least four effects on the state-led adjustment model. First, the absence of central levers, such as strike funds, made centralized control of labor disputes difficult. Second, because mobilization was separated from bargaining, employers were reluctant to include the unions as regular interlocutors. Third, competitive unionism impeded the development of a class coalition capable of confronting industry and forcing it to modernize. Fourth, labor's contribution to the adjustment process would be far more reactive than constructive. This combination complicated efforts to alter the triangle of business, labor, and state relations.

The Chimera of Growth with Inclusion

When postwar planners looked at the American model, they found not only industries with enormous technological capacity. They saw an economy in which wage increases fueled mass consumption that stimulated demand for mass production goods.[66] In spite of attempts to modernize the French economy, planners had only mixed success in combining growth with labor's input.

Governments had tools to manage the economy by the early 1950s.[67] Policy makers after 1945 developed national accounting methods through the Institut National de la Statistique et des Etudes Economiques (INSEE), medium-term forecasting capacity, and statistical modeling techniques that enabled a better understanding of economic activity.[68] Growth and productivity rates were high in the 1950s and 1960s, but increases in spending (direct and social wage included) were irregular.[69] Real wages rose consistently in these two decades, but year-to-year fluctuation was considerable. This irregularity had two sources—the late adoption of demand management tools by the state and relations between labor and management at the workplace. The result was considerable fluctuation in purchasing power in the 1950s and early 1960s.

Postwar demand management was both weak and state-directed. Because of state control of the financial sector, government officials did not need increases in demand to stimulate investment. Carré, Dubois, and Malinvaud estimate that half of economic growth throughout the first two decades after the war resulted from peasants moving off the land.[70] The social security system was less progressive than its counterparts in other countries. Spending was also concentrated in family allowances and pensions that did not fluctuate in cyclical downturns, especially given the late development of national unemployment compensation.

Trade unions participated only peripherally in policy making. Union members sat on the Economic and Social Council after 1946. They also administered social security funds after that date and unemployment funds after 1958. They participated in tripartite bodies that administered training programs after 1971. They had access to departmental labor inspectors. However, this inclusion was limited either to inconsequential decision making or implementation of decisions made elsewhere. Consequently, the policy networks typical of a strong state developed in France in the absence of a labor component. Organized labor could influence policy only from outside.

Until the mid-1980s, attempts to coordinate labor policy with adjustments in the product market failed because of a lack of labor market participation. By the 1960s, the state sought to coordinate labor and product market adjustment, yet was unwilling or unable to pursue an inclusionary strategy to forge that combination. It did not offer sufficient inducements to encourage organized labor and management to cooperate. The government refused to alienate traditional political constituencies.

The most celebrated attempt to forge that linkage took place with the "new society" program of Prime Minister Jacques Chaban-Delmas. After the turmoil of May 1968, the prime minister for the newly elected President Pompidou sought to "institutionalize" the labor market to provide the stability that would be consistent with the industrial policy of national champions. The government offered to develop aggregate remuneration levels within the planning process. Employers and unions, free of state interference, would bargain over how they were distributed. To sweeten the procedure, cost-of-living adjustments would keep wages ahead of inflation. In return, unions would agree to conciliation clauses making strikes more difficult to initiate. This would be possible, so the argument went, because employers would show new respect for union organization, and unions would participate in the process.

Labor policy coordination included the initiatives on training described above—the 1970 collective bargaining agreement and the 1971 law. Although less concerned with employment protection, the Ministry of Labor encouraged unions and employers to sign an agreement that requested that all layoffs be authorized by the departmental labor inspector. The linchpin was the incomes policy that would stabilize earnings and firm-level expectations.

The "new contractualism" failed to introduce new linkages between labor and product market adjustment. Although the number of collective bargaining agreements increased between 1969 and 1971, employers remained hostile to the entrenchment of union power on the shop floor (in spite of reforms after May 1968). Moreover, union opposition to the strike clauses prevented the push toward long-term agreements sought by the government. Finally, state actors were suspicious of private agreements that might contravene public policy. Including unions even in routine collective bargaining would have required different shop-floor relations between employers and unions and a labor movement with a different historical legacy.

At the root of this disjunction was the organization of the French firm. Postwar management was not prepared to share power with labor, and labor was not prepared to concede the "divine right of management."[71] Although there was controversy within the CNPF over political tactics vis-à-vis labor and the state, traditional elements maintained the upper hand until the late 1960s. Organized business was reluctant to dilute property rights or consider legalizing union sections at the workplace—the latter veto justified as defending the right of property. The CNPF worked tirelessly to dilute the competence of works councils. Along with other business associations, it opposed both the ECSC and the Common Market. Employers sought to enhance productivity but were reluctant to share the proceeds of that new efficiency.[72]

By the 1960s, an impending shortage of workers pressured employers into rethinking how they ran their factories. Bloch-Lainé's essay, *Pour une réforme de l'entreprise*, served as a catalyst for the modernists within employer circles. These forces included Catholic activists in the Centre des Jeunes Patrons who wanted French business to compete in open markets and negotiate with labor, explicitly promoting the union section.[73] They sought to change the structure of the firm to increase efficiency, even if this placed state officials and worker representatives on the board of directors! Their advocates were defeated in 1965, excluded from the executive council of the CNPF.

This struggle over control of the firm was not even addressed until the aftermath of May 1968. The Gaullist strategy of national champions increased firm size and provided a recruitment trajectory for a new generation of business leaders who tended to come more from the *grandes écoles* than from owners' families.[74] These managers played a more important role in the trade associations. The management structure of the firm changed from family-controlled holding companies to multidivisional corporations with strong marketing divisions.[75] By the early 1970s, employers began to see more utility in negotiating with organized labor, partly as a result of "new thinking" within the CNPF, partly as a result of state strong-arming.[76]

The assumptions justifying labor inclusion had changed by the 1980s. The internationalization of the economy sharply limited the possibility that political action could effectively master economic forces. The more international the economy became, the more state planning consisted of implementing market-driven imperatives rather than forging social change. As economic power devolved to local firms, central control became more and more counterproductive. Labor was eventually incorporated into this process, but the nature of its inclusion has frustrated the original aspirations of the labor movement. To paraphrase Schattschneider, bias had been mobilized in the wrong direction—at least for French unions.

Conclusion

The state-led adjustment model developed along a trajectory different from that envisioned by the Monnet group. Far from an arena to mobilize capital and labor together, it became a vehicle to prod business in product markets while pleading with firms to change their labor market reflexes. This strong state in France had many weaknesses.

The failure to integrate labor into the adjustment model left only unorthodox methods whereby labor could articulate its demands, and labor protest remained a regular tool in the union repertory. Given the ideological heritage, the oppositional response came effortlessly for unions. Labor mobilization met with an even more natural intransigency by many companies. It is not surprising that "modernizing the firm" became a perpetual goal of successive governments.

The longer labor was excluded, however, the more difficult it became to include unions in policy making. French companies developed a hostility to labor participation. Even mass producers marginalized unions from either wage setting or personnel development. They developed a "bargain basement Fordism"—based on low wages, an emaciated system of industrial relations, and hostile relations between assemblers and subcontractors to maintain that price advantage—that held together as long as the international economic climate was smooth and the nation-state was the locus of economic policy.[77]

The modernizers' insistence on trade liberalization as a way to defeat Malthusianism had potentially pernicious side effects. While state intervention provided a growth-oriented economic policy, it was hardly compatible with an open international environment. Yet that open environment was essential in whipping the recalcitrant French capitalists into shape. The more open French borders became, the more the interventionist apparatus suffered from diminishing returns. If capital were mobile, for instance, companies might look elsewhere for financing. Thus the implicit wager: if the modernizers succeeded in creating a healthy French capitalism, they would reduce the power of the interventionist state.

French development depended on a significant degree of autonomy for national policy. The state made financial allocations to large firms in hopes that they would consolidate market shares while maintaining intact a significant small business sector for political stability. Rapid economic growth, payoffs to traditional sectors, and even acquiescence in wage demands to quell social protest built inflation into the state-led model. Only timely devaluations of the franc restored the competitiveness of French products.

The tension between open borders and the national model was acute, given the position of labor in the French economy. Policy leaders could not rely on a tradition of good will to negotiate new terms of adjustment. Having failed to articulate and coordinate labor market and product market strategies, firms could rely on few indigenous sources of wisdom, and managers tended to search abroad for methods to restore competitiveness. Labor was hardest hit: having mounted protest for meager returns with incipient mass production industry, the unions had difficulty mobilizing themselves and negotiating over quality and participation—the keys to competitive success in the 1980s. The following chapters explore these dilemmas in the steel industry.

THREE

The Postwar Steel Industry

The iron and steel industry has long been strategically important to the French state and economy. It was key to a strong armaments industry, and it belonged to geopolitical *grandeur*. Prussia's seizure of part of the industry in 1870 came as no surprise. By the twentieth century, the steel industry provided the necessary materials for many consumer durables. An efficient and productive steel industry seemed essential for postwar reconstruction, economic prosperity, and national power.

The most striking feature of the steel industry in postwar France is its rapid rise to prominence and its even more dramatic decline. In 1945, some plants and equipment still dated from the Prussian occupation. By 1961, an article in *Paris Match* referred to Lorraine as "Texas, France," so awed was the author at the frontier potential of the region.[1] Steel production in Europe was booming, and Lorraine was in the heart of a Europe in the process of integrating its potential within the European Coal and Steel Community (ECSC). Coal and steel seemed to represent the essence of a revitalized continent and nation. Economic growth in the region was high, and it attracted immigrants hoping to make their fortune. Only five years later, however, the steel industry threatened Lorraine with bankruptcies and employment losses.

The rapid rise and fall of French steel reflects how the struggle among the state, industrialists, and steel workers unfolded in the unusual circumstances of the immediate postwar period. Modernizers sought to reform an industry characterized by small plants, inefficient production, and a trade association designed to protect the interests of the Malthusians. Steel employers were suspicious of such modernization efforts. They fought tooth and nail to protect their profits, preserve their autonomy from state intervention, and maintain paternalistic control over their employees. The unions sought to replace the backward capitalists. They demanded that the industry be nationalized and the economic surplus distributed more equitably. At the company level, they attempted to organize the very workers whom the employers hoped to keep compliant. The result was a set of compromises and changes that nobody had fully anticipated at the war's end.

This chapter explores the economic and social position of the postwar French steel industry. State-led strategies to increase efficiency through the ECSC and internal consolidation of industry assets were partial successes. The ECSC set constraints on French producers but also protected them from price-cutting by European competitors. Assets were consolidated into four major companies. Yet efficiency failed to increase enough to gain international competitiveness. These uneven outcomes of "arm's length intervention" in product market policy decisions reflected both effective organization by French industrialists and the weaknesses of state control. In return for partial cooperation on industrial policy, the state tolerated paternalistic practices in labor market management. These compromises smoothed relations between the state and steel industrialists as long as domestic demand rose, trade relations remained stable, and workers lacked the means to reverse their exclusion. Over the long term, though, they hampered modernization, perpetuated inefficiency, generated labor opposition, and increased incentives for state intervention.

The Fragile State of Steelmaking during Reconstruction

After the German occupation, modernizers, industrialists, and workers all wished to revive the steel industry in the interest of rebuilding the nation. Everyone agreed that France was not fully competitive because steel was produced in small plants that used inefficient machinery and low-skill manufacturing techniques. These characteristics of the industry, however, had important political consequences. The modernizing state wanted to streamline the process and increase efficiency so that national supply could keep up with demand during an expansionary phase. Industrialists, by contrast, had a vested interest in maintaining their modes of production and wanted protection from the rigors of international competition.

In the postwar period, integrated mills combined the many steps of steel manufacturing. Despite dramatic technological changes after 1950, the three production steps remained constant. First, a blast furnace mixes iron ore, coke, and limestone to remove oxygen from the ore and create pig iron. Second, a steel converter heats the pig iron to a temperature sufficient to reduce the carbon content to under 1 percent to produce crude steel. Adding alloying elements at this stage alters the physical properties of the crude steel and transforms it into various specialty steels. Third, rolling mills shape steel ingots into the forms desired by downstream industries. This is done by using rollers to squeeze and pull the product to a desired shape and thickness. Finished long products—rails, sectionals, pilings, bars, and wire rods—constitute a critical engineering product for construction. Finished flat products—sheet, strip products, and plate—are used for vehicles, consumer appliances, and ships' hulls.

Steel production is generally considered to be capital-intensive and requires some of the largest-scale facilities of any industry. Over the postwar period, a premium was put on extending economies of scale. Technical innovation in enlarging blast furnaces and speeding up rolling mills increased several times over between 1950 and 1980. The shift from open hearth to oxygen steel conversion both saved time and reduced energy costs. By transforming molten steel into semifinished products ready for immediate rolling, continuous casting eliminated the ingot stage, thereby forgoing costly heating procedures. By the 1970s, new technologies permitted small-scale production for local markets in minimills via electric steel converters that used scrap to produce pig iron—thereby eliminating the blast furnace—and continuous casting machines. These minimills reduced energy and capital costs, although they produced mostly simple rolled products—bars and wire—used for local markets. They were less important in the French than in the Italian or U.S. markets (for reasons explored below), and the large integrated facility that combined the three steps of the steelmaking process remained the ideal for planners until the 1980s.[2]

In 1950 French steelmaking was still characterized by small plants and technology from the pre-1914 period. Ownership was both factionalized and highly rivalrous. No new plants had been built in the interwar period. Instead, the industry felt lucky to have confiscated German equipment in Lorraine from the Treaty of Versailles and from the return of territories. Plant owners also relied on production strategies developed during the German occupation, in the belief that the Germans would be willing to trade rich Ruhr coal for Lorraine iron ore.

French steel production in the early postwar period was also extremely labor-intensive. Unsophisticated technology and fluctuating business cycles created a demand for two types of workers. To ensure uniform production, managers needed a core group of skilled workers who had developed rule-of-thumb procedures through long experience at a particular plant. They also needed a secondary labor market of docile, dependable, yet "disposable," workers to absorb fluctuations in demand.

Paternalism complemented these production strategies. Because of labor-intensive production, owners found it important to retain the most skilled workers. Unsophisticated technology required on-the-job skills to guarantee uniform quality and to ensure the perpetuation of rule-of-thumb procedures. On the other hand, firms needed a secondary labor market because of the sensitivity of orders to the business cycle.

The state had targeted steel as a key sector in the First Modernization Plan in 1946, and the industry received considerable financial resources after that time.[3] Planners channeled Marshall Plan monies to the steel industry. It also benefited from infrastructure subsidies. The canal linking the Moselle and Rhine created an inland waterway in Lorraine; it was funded as compensation to the steel producers for the ECSC.[4] Likewise, rail links were modernized and worker housing was built. During the 1950s, low-interest loans and state guarantees accounted for almost 40 percent of new investments.[5] Nationalized banks extended loans and guarantees. Until companies began to fail, state funding came with few strings attached.

The steel men, however, did not trust the modernizers. They had been suspicious of state interference since the Liberation. The first concern of the steelmakers in 1945 was control of assets. The National Council of the Resistance had called for nationalization of the steel industry as part of a recovery program. The steel men (*maîtres de forge*) were viewed with jaundiced eyes by the Left because of their political intrigue between the wars.[6] Nationalization seems to have been delayed, however, because of the competing exigencies of constitution writing and political jockeying after the Liberation.[7] Afterward, there was neither a sustainable political coalition for nationalization nor much enthusiasm for it among modernizers. Periodically, however, the Communists introduced bills onto the floor of the National Assembly, and nationalization remained on the lips of most unionists. Given the threat of nationalization, steelmakers could not risk a clearly antagonistic relationship with the state.[8] These twinges of doubt over their continuing ownership undoubtedly reinforced an entrepreneurial timidity among the steelmakers.

Industrialists also distrusted the state's capacity to interfere in product markets. The governance of the industry before the war had been undertaken in the name of market stability. When in 1945 Parisian technocrats wanted to force property owners to consolidate their assets, steelmakers refused to shut unproductive plants. They gave only modest support in the sectoral committee of the Plan.[9] On the whole, industrialists sought to preserve their autonomy and to minimize state interference.[10]

Given the backwardness of the industry, however, French steel producers wanted the state to protect them from foreign competition. The industry had a long history of protection. It had participated actively in the International Steel Cartel in the 1920s and 1930s. After the war, the steel producers of the Nord wanted to modernize to gain competitiveness. The Lorraine firms, meanwhile, hoped to continue such market constraints. Since they produced 70 percent of French steel in 1950 (figure 3.1), they dominated the industry association, the CSSF. Thus, the CSSF opposed the state over the formation of the ECSC. During negotiations over the Treaty of Paris in 1950–1951, steelmakers refused to open their books for Monnet on the grounds that it would be an infringement of industrial secrecy.[11]

The modernizers saw the ECSC as a way to manage both external and internal political problems. The Community attempted to settle once and for all the *Ruhrpolitik*—the massive German advantage in heavy industry that facilitated expansionary foreign policy.[12] By forging cooperation with Germany, entangling it in a web of competitive dependency, and setting rules for production and trade of the means of aggression, the Treaty of Paris would maintain Germany in the world of nations. To Monnet and French modernizers, the ECSC would be coupled with decartelization of the German *Konzerns* to ensure the integration of a peaceful Germany into Europe. According to some accounts, Monnet had even grander plans. Frances Lynch argues that the ECSC was part of a larger attempt that began with the First Plan to assert the

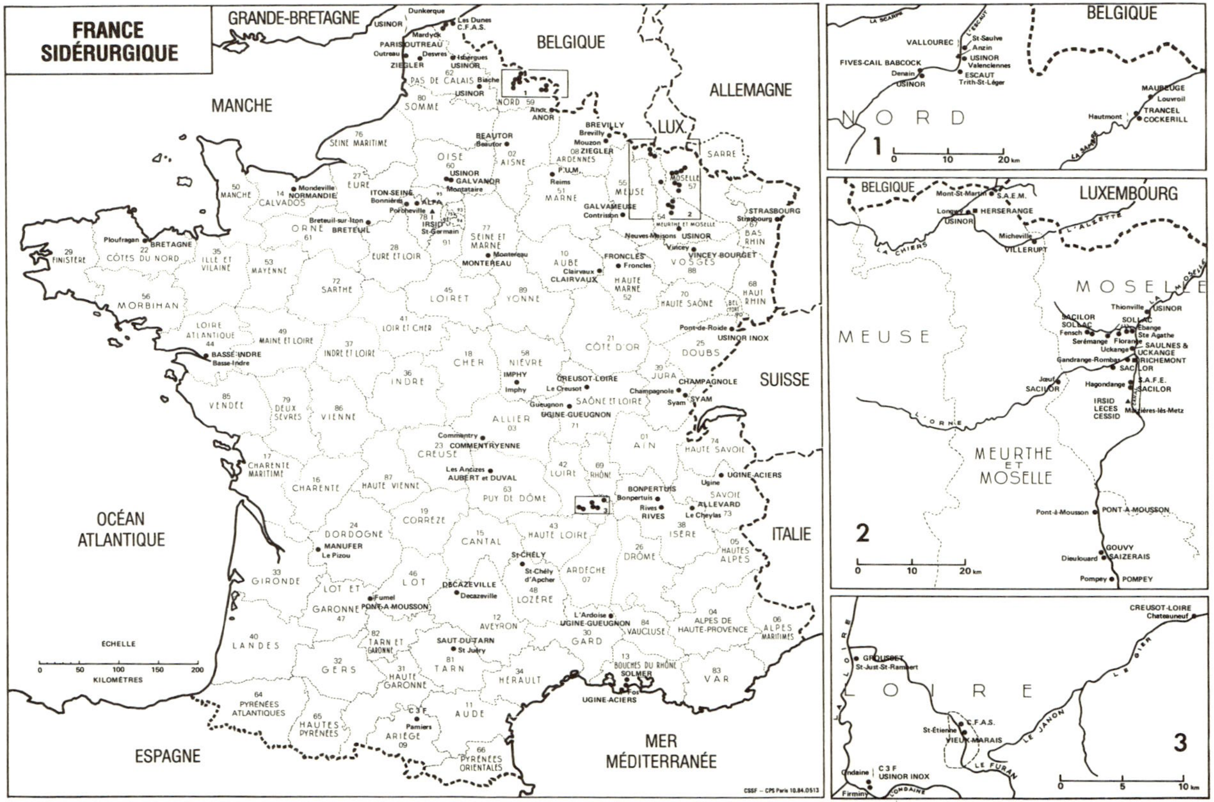

Figure 3.1 The Geography of the French Steel Industry

dominance of French heavy industry by guaranteeing access to the Ruhr coalfields.[13] To French producers and the steel unions, however, the treaty promised to bring an invasion of German products into French markets.

The European Coal and Steel Community and the Development of French Steel

The ECSC proved much more nuanced than any of these scenarios. German industrialists outmaneuvered the decartelizers among the allies. By the mid-1950s, steel capital was more concentrated in Germany than in France. Access to German coal was not reciprocated by the importation of French iron ore. Still, the transnational threat was negligible, for each industry served a national market. While imports among Community members increased throughout the postwar period, they remained less than the world average.[14] Few cross-border mergers took place.

This anomaly—recartelization of the Ruhr, yet continued market protection for French producers—was the result of the organization of the ECSC. Monnet claimed that it would be a supranational organization not directly responsible to national governments. The ECSC High Authority maintained relations with steel firms unmediated by state authorities. Monnet envisioned executive privilege as facilitating the development of a European economy along the lines he thought most appropriate—a combination of European indicative planning and free markets.[15] The thrust of Community organization was more *dirigiste* than liberal, however, and led more by the High Authority than by companies themselves.

Monnet's vision existed only on paper. The Community's executive was hamstrung by the competing demands of steel producers. While Monnet sought political authority to open markets, producers managed to reestablish many of the pre-1950 cartels that had restricted competition. While the High Authority sought intra-Community cooperation on investment, production facilities, and labor practices, national norms remained in place.

Supranationalism was not achieved because of contining oligopolistic industry structures. Because of fears from French producers about an impending recession, the Community did not set up an unrestricted free market for steel products. Instead, base-point pricing (BPP) guaranteeing that local producers could not be undersold set minimum prices in regional markets, and these markets did not transcend national borders. (In the United States, antitrust laws restricted this practice as a restraint on trade.) Pricing points could be subject to large-volume discounts, thereby permitting flexibility in its operation. With local firms agreeing to common prices and with only those prices prevailing, steel consumers would always know what they were to pay for individual products. Such transparency discouraged discrimination, but it also settled price competition within a given region. By fixing prices locally and enabling firms (not governments) to monitor them, BPP contributed to a *national* cartelization of the European market. National cartels, in turn, undermined Monnet's supranationalism.[16]

Governments were also unwilling to relinquish authority over their steel industries. The most basic executive policy tools of the ECSC were the abilities to tax and borrow. The Treaty of Paris provided for a levy (up to 1 percent) on all steel products, which could be used for investment, research, or worker adjustment. The tax became a political football between the High Authority and national governments represented in the Council of Ministers. The rate was lowered after the mid-1950s. The Community did allocate its revenues for worker housing, retraining, and relocating redundant workers. Still, the monies passed through the hands of national authorities who set the purposes for which total funding would be used.

The constraints on the ECSC policy levers were evident from the beginning. In 1954, Monnet negotiated a loan from the Eisenhower administration and later floated international bond issues both to intervene selectively in individual investment projects and to bolster the image of the High Authority.[17] Yet it was insufficient to influence investment choices. Lenders preferred the financing of steel investments outside the auspices of the ECSC.

Far from freeing trade among the Europe of Six, the ECSC resulted in "managed markets," and the national industries themselves managed them. In defiance of the High Authority, European producers organized export cartels that allowed high prices for third country markets (mainly the United States) to compensate for discounts in national markets. Thus, each nation's industry enjoyed the freedom to pursue a national strategy.

Paper *dirigisme*, however, was important for its implicit powers. Aside from unfettered access to individual firms, the High Authority (after 1967, the Commission) had authority to declare a "manifest crisis" within the Community and use special powers to organize the market by commanding prices and reducing production. As we will see in chapter 7, these powers helped the French steel industry adjust after 1974.

In spite of Monnet's intentions, the ECSC enabled the French industry to pursue nationally specific strategies until the 1970s. This kept the Community from setting the agenda in French steel policy until the 1970s. In market terms, the ECSC protected the French market for its high-cost production. In political terms, its directives later shielded decision makers from the public fury inflamed by announced cutbacks.

If modernizers and industrialists were divided over relations between the national steel industry and the European Coal and Steel Community, they seemed united over the need to increase internal economies of scale. Company managers, Economic Recovery Program administrators, and French planners agreed that mergers were needed to consolidate assets. Concentrations took place in the late 1940s and early 1950s (see figure 3.2). In Lorraine, De Wendel et Cie. absorbed Les Petits Fils de François de Wendel in 1950, reuniting assets separated by the German occupation after 1871. Sidélor (Union Sidérurgique Lorraine) was created by the 1950 merger of three smaller producers. Lorraine-Escaut was formed from three Longwy firms in 1953. In the Nord, Denain-Anzin and Nord-Est combined capacities to form Usinor (Union Sidérurgique du Nord de la France) in 1948. These four firms accounted for 54 percent of French crude steel output in 1953.[18]

Concentration of assets did not bring economies of scale, however. Plants remained small in the 1950s: by 1960 only two integrated plants produced 1 million metric tons (MT) of steel, the internationally recognized threshold of production efficiency. The two plants that benefited from Marshall Plan funding and American technology—wide-band strip plants at Usinor-Denain and Sollac—did not meet the standards of newly designed mills. Capacity specifications for the rolling mills dated from the 1920s.[19]

Sollac epitomized these disparaties in concentration. No Lorraine producer had the capital to meet the specifications for Marshall Plan steel money—1 MT of capacity. Therefore, a cooperative was formed under the name of Sollac (Société Lorraine de Laminage Continu), headed by de Wendel, which owned 47.5 percent. In amounts equivalent to ownership shares, each member delivered the amount of pig iron or semifinished products for conversion to steel or rolling into finished sheet. Being dependent on several different owners, Sollac could not

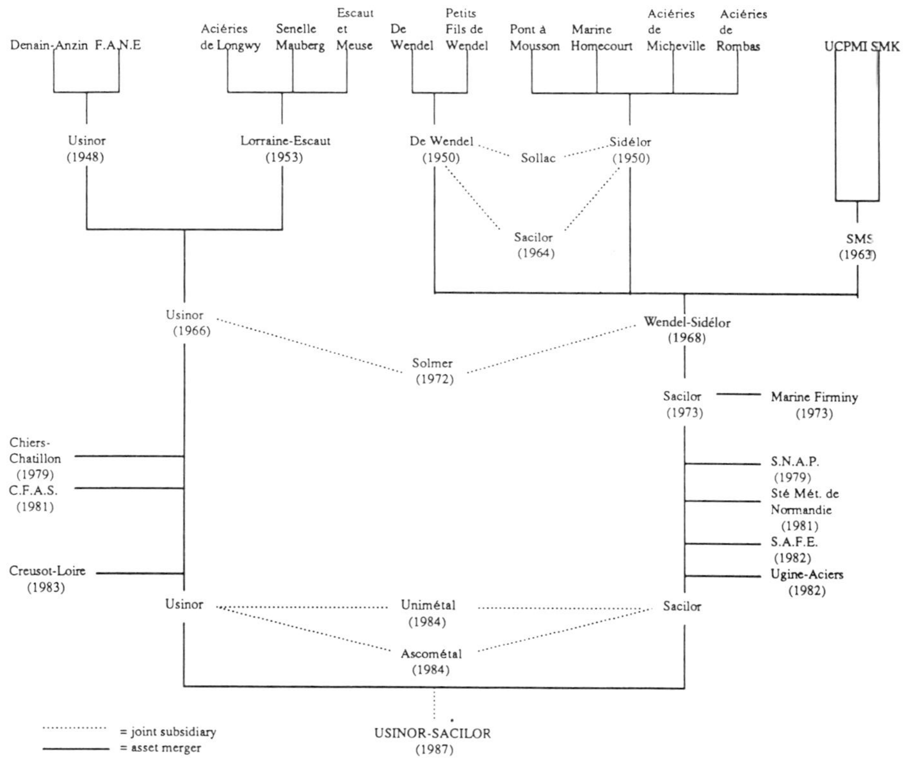

Denain-Anzin F.A.N.E
Aciéries de Longwy
Senelle Mauberg
Escaut et Meuse
De Wendel
Petits Fils de Wendel
Pont à Mousson
Marine Homecourt
Aciéries de Micheville
Aciéries de Rombas
UCPMI SMK
Usinor (1948)
Lorraine-Escaut (1953)
De Wendel (1950)
Sollac
Sidélor (1950)
Sacilor (1964)
SMS (1963)
Usinor (1966)
Wendel-Sidélor (1968)
Solmer (1972)
Sacilor (1973)
Marine Firminy (1973)
Chiers-Chatillon (1979)
C.F.A.S. (1981)
S.N.A.P. (1979)
Sté Mét. de Normandie (1981)
S.A.F.E. (1982)
Ugine-Aciers (1982)
Creusot-Loire (1983)
Usinor
Unimétal (1984)
Sacilor
Ascométal (1984)
= joint subsidiary
= asset merger
USINOR-SACILOR (1987)

move aggressively in its production choices and failed to maximize its potential.

Production and investment in the French industry had a mixed record in the 1950s. On the one hand, producers needed to rebuild plants and provide the steel needed for reconstruction. Demand outstripped supply in the international steel market, especially during the Korean War boom, and surplus capacity was not considered an issue until the early 1960s. The CSSF constantly touted every increase in tonnages as an element of national pride. On the other hand, as can be seen in table 3.1, growth in tonnages was slower in France than in the two dynamic steel producers of the ECSC—Germany and Italy.

The French were not really so productive as they claimed. While labor productivity figures are notoriously unreliable, those published by the Planning Commission showed convincingly that France lagged behind other producers in efficiency (see table 3.2). France was a high-cost country throughout the 1950s and 1960s, compared to its ECSC rivals.[20] France's poor performance had at least five causes.

First, capacity use remained low, especially for rolling mills, because of the failure to retire old plants as new units were built. Small-scale, family-owned business used this capacity as a hedge against overly optimistic forecasts of future markets. Producers neither rationalized production within existing plants nor, with the exception of Sollac, did they build new plants in Lorraine in the 1950s. Sidélor grouped together assets, but its members continued to produce similar products in simi-

Table 3.1 Crude Steel Production in France, Germany, and Italy, 1952–1966
(in millions of tons)

	1952	1954	1956	1958	1960	1962	1964	1966
France	10.9	10.6	13.4	14.6	17.3	17.2	19.8	19.6
Germany	18.6	20.2	26.6	26.3	34.1	32.6	37.3	35.3
Italy	3.6	4.3	6.1	6.4	8.5	9.8	10.2	13.6

	$\frac{1960–1952}{1952}$	$\frac{1966–1960}{1960}$	$\frac{1966–1952}{1952}$
France	59%	13%	80%
Germany	83	4	90
Italy	136	60	278

Source: Eurostat, *Iron and Steel: 1952–1982.*

Table 3.2 Number of Hours to Produce a Ton of Steel, 1957–1966

	1957	1958	1959	1960	1961	1962	1963	1964	1965	1966
France	24.8	23.9	22.8	21.0	20.9	20.8	20.3	18.3	17.7	16.4
Germany	16.4	16.8	15.5	14.9	14.9	14.6	14.4	13.6	13.1	12.7
Belgium	19.7	20.8	19.8	18.0	18.3	17.9	17.4	15.4	14.0	12.5
Luxembourg	12.6	13.0	11.6	11.1	11.1	11.2	11.1	10.0	9.9	10.0
Netherlands	19.0	16.3	14.8	14.2	14.4	14.3	13.1	12.6	11.4	10.8
Italy	20.4	20.6	18.2	16.1	14.2	14.9	14.6	14.1	11.3	10.0

Source: French Planning Commission, unpublished data, various years.

larly outdated facilities. The organization of De Wendel left the family in control: fearing nationalization, it positioned its assets so as to take advantage of their financial and not their industrial complementarity. Between 1947 and 1957, the company focused exclusively on rebuilding existing facilities, with major investments going to Sollac, and it invested significantly only in iron making—not steel making—toward the end of the 1950s.

These family businesses also used managerial strategies that compounded the small plant problem. Companies feared losing clients more than they attempted to seek new ones. Consequently, they insisted on manufacturing a wider range of products than was economically efficient. According to a former executive for Usinor: "One felt obliged to produce all dimensions of flat products . . . in order to prevent clients from going next door. That led to unceasing adjustments of the mill and disastrous productivity."[21] Such reasoning brought overinvestment for underused facilities.

Second, mergers were designed to consolidate assets rather than facilitate production efficiencies. The "Lorraine strategy" grouped together assets to produce a similar range of products and integrate upstream to include iron ore mining—regardless of ancillary costs. The formation of Sidélor was a prime example of a strategy that failed to maximize productive potential. The Cie. de Micheville in Villerupt was geographically better placed to merge with the companies that formed Lorraine-Escaut than with Sidélor. This would have created a line of production from Longwy through Villerupt to Thionville. Likewise, the major steel facilities of the Société Lorraine des Aciéries de Rombas abutted those of the De Wendel holdings in Joeuf, yet they were integrated into Sidélor. The ill-fated merger that created Sidélor had its origins in the consolidation of iron mining assets.

Third, a production strategy based on using Lorraine iron ore crippled the industry. The quality of Lorraine ore was poor, containing only 30 percent iron, while Swedish ore contained 60 percent. European competitors later used African, Latin American, and Australian ore that surpassed 65 percent in iron content. Sintering and enriching of the Lorraine ore was expensive. Moreover, because the ore contained a high phosphorous level, standard open-hearth converters produced inferior steel. In the 1890s the open-hearth process was adapted to produce steel in so-called Thomas converters. The resulting steel was cheaper to produce than simple open-hearth steel because it required less personnel, but it was more brittle due to the remaining levels of nitrogen. The Lorraine industry could have enjoyed the benefits of local ore if the mining companies had not "topped off" the reserves, mining the richest deposits first for export.[22] Therefore, successes in converting an increasingly poor quality of ore diverted resources from other uses.

Fourth, French producers failed to invest sufficiently in new technology. Steel conversion lagged behind alternate technologies in the early 1960s. Oxygen steelmaking was partially delayed in Lorraine by the use of oxygen blowers in Thomas converters, which increased quality and productivity, but did not approach the cost advantage of oxygen converters used elsewhere. Sollac used the modified Thomas conversion process until the 1970s—long after the open-hearth process had become internationally uncompetitive. When investments were made, the firms could not use new equipment for optimal productivity. (Ironically, by some estimates, they invested more per ton of capacity than their European rivals.)[23] At the Lorraine-Escaut plant in Thionville, for instance, the blast furnace was technically among the most productive in France, but it could not be used to full capacity because of the size of steel converters and rolling mills at the site.

Fifth, steel production was divorced from steel processing activities, involving a greater value added.[24] The German firm Thyssen, for instance, streamlined production from coal through steel to investment and consumer goods. Some degree of downstream manufacturing did take place in the groups organized by the steel holding companies. However, they were not organized to maximize coordination. Moreover, manufacturing activities were limited to pipes and basic transformation, such as wire netting and cut nails. The major producers made only crude steel products and left the bulk of specialty steelmaking to producers in central France. This reduced the profits of the majors and made specialty steel production a high-cost venture, for it lacked the scale that even Lorraine or Nord producers enjoyed. Ironically, the Loire producers had downstream diversification but not the economies of scale to compete effectively.

Despite the economic problems facing them, producers rode buoyant markets during the 1950s: production, profits, and share prices all rose. As a critical component of mass production industries, steel remained in demand. Profits enabled the industry to reduce private indebtedness in the late 1950s by substituting state credit, making the companies more dependent on state borrowing. Workers were recruited enthusiastically. Usinor's coastal plant in Dunkerque was finished in 1963, with one of the best productivity rates in Europe. Lorraine and the Nord were two points on an axis of an industrializing Europe.

Nonetheless, French steel plants were small, poorly coordinated, based on increasingly uncompetitive iron ore and increasingly outdated technologies. Because of the small size of French steel companies by international standards, the expense of new equipment became a heavy burden. Low productivity brought profit margins insufficient to generate investment funds. Companies were more exposed to market fluctuations because of the lack of greater value-added activities.

The Politics of Early Industrial Policy

Policy makers in the Ministries of Finance and Industry were well aware of the industry's predicament. Yet the modernizers' enthusiasm did not translate easily into state policy to reorient the industry. They made major concessions to the industrialists on both the ECSC and consolidation of assets that undercut their efforts to increase steelmaking efficiency. Industrial policy was constrained by bureaucratic rivalry, the producers' power to protect their interests, and the state's unwillingness to tinker with what seemed ironically to be a success story.

Industrial policy in steel was characterized by bureaucratic rivalry. Jack Hayward suggests that considerable tension existed among the MoF, the MoI, and the Planning Commission over the appropriate direction for steel policy. Apparently the CSSF used its contacts in the Metallurgical Industries Division of the MoI to temper the "expansionist" Planning Commission and the "conservative" MoF.[25] Consequently, the industry controlled the policy agenda. The modernization commissions of the Plan relied exclusively on the industry's interpretation of the market. Conflict existed over various targets, but only on the basis of data provided by the companies.[26]

State price controls only partially upset this cozy business-state relationship. According to the ECSC, firms could set their own base prices. However, the MoF pursued a strategy of monitoring the overall price level and influencing firm behavior through exceptions. This included steel. The MoF used its influence on steel users by constraining prices, and those users then refused to pay higher base prices to the steel producers. Perhaps more important, the MoF made voluntary "compliance" by the steel producers a condition of investment aids. The High Authority fought continuously with French governments to raise prices for the French market but to no avail.[27] Price controls, as suggested in chapter 2, were also levers to influence production choices. Throughout the 1950s, the prices of Thomas steel products were kept below those of their open-hearth counterparts. Given the dependence of the Lorraine producers on Thomas steel, they were disproportionately hit by controls. This tool was not combined with other production incentives, though, so the thoroughness of this policy remains uncertain.

The MoF paid for price controls with subsidized credit from the FDES. This quid pro quo of credit availability for price controls irked the companies. The industry bitterly contested controls on all steel products, later blaming market failure to below-market prices.[28] It attempted to buck the price controls in 1956 with a 4 percent price rise, over the MoF's objections. In response, the government raised the price of imported coal, reduced interest subsidies, and lowered export subsidies, effectively offsetting the industry's price hike. The industry rescinded its price increases.[29] In 1961, another attempt was made to raise prices by 2.5 percent for a fraction of steel products, and the minister of finance severed official contact with the CSSF for two years.[30] Consequently, steel prices averaged 10–15 percent lower than those in Germany or Belgium, weakening company finances.

Business-state collaboration facilitated the allocation of credit within the industry. The MoF worked with the CSSF to organize the Groupement des Industries Sidérurgiques (GIS), which issued bonds guaranteed by the industry. The ministry both granted the GIS access to the tightly regulated capital market and gave it the time on its "calendar" to offer the bond issue. Collective borrowing brought better terms than any individual firm could get in the bond market. While it increased the overall availability of privately raised credit, small firms benefited disproportionately. GIS issues were responsible for a third of all investment money raised by the industry.[31]

The organization of the GIS reinforced the CSSF. Because the GIS board vetted individual investment projects for financial soundness, it could exert influence over the industry. Composed of industry members, the GIS board was headed after 1952 by Jacques Ferry, then vice-president of the CSSF. Once the GIS issued borrowings, its dossiers were examined by the MoF and the Steel Directorate of the MoI, both of whom could raise objections. Thus, the mechanisms of credit allocation brought a pyramid of authorizations that placed the CSSF via the GIS in a position to influence industry politics. According to John McArthur and Bruce Scott, "In steel the financing of major investments was anything but a confidential affair between a company and its banker: it was a matter of common knowledge in the industry and a matter on which various industry and State sources had time to 'reason together.' "[32]

French steelmakers both needed support from the state and sought to guard their autonomy. The trade association performed the balancing act. A government decree had made CSSF membership compulsory for all steel firms. The CSSF allocated raw materials in the immediate postwar period and ran the cartels that organized French production until the ECSC. Afterward, it processed most sales of steel products, so it had its own sources of data.[33]

The CSSF worked with the state to govern steel politics. Internal cohesion gave the trade association external strength. However, that strength was built on a hands-off policy toward its members. Until 1963, with the election of Jacques Ferry, presidents came from the steel firms. Leaders sought to build consensus within the organization and not pursue policies that could marginalize individual producers.

"Arm's length interventionism" characterized business-state relations until the mid-1960s. There were tensions over price controls, but the state restrained its interventionist powers. It accepted the industry's view of the market and played the role of facilitator. The industry had temporarily constrained the modernizers from Paris and Brussels.

The Social Basis of Traditional Industry

Steel employers also tried to regulate the behavior of their employees. As in their relations to the state, they sought to retain and solidify managerial autonomy. But such independence atrophied for reasons over which employers had little control.

The effort to attach workers to the firm through nonsalaried incentives has been an employer tool under various guises at different periods of industrial development. According to Donald Reid, paternalism implies that the firm "owed workers not just their wages, but solicitude and guidance as well."[34] As a system of social relations, paternalism was not unique to France; "company towns" have existed everywhere.[35] The French steel regions (especially Lorraine) differed from analogous industrial areas in France and in other countries because of the tenacity of these authority relations: paternalism did not evaporate until the 1960s. It shriveled in the major U.S. economic sectors in the 1930s, as industrial unionism developed and as the federal government assumed a larger role in forming labor policy. German paternalism was killed by a combination of the Allied forces and codetermination. French paternalism limped into the postwar period because firms remained small, unions were sufficiently weak, and the state was reluctant to intervene in labor relations.

The relationship between employer and employee in the early 1950s in Lorraine displayed many of the "feudal" characteristics of the prewar period. Immediately after the war, however, this was not evident. The purge committees (*comités d'épuration*) successfully struck fear into the hearts of managers who had continued working during the German occupation. By 1946, industrial activity was back to normal. The major strikes in 1947–1948 affected steel as well as the rest of the country. They were defeated by a combination of arrests, requisitions, and state intransigence, since steel production was so critical to reconstruction. Consequently, union mobilization died down. The CGT was particularly hard hit. According to police reports, CGT membership dropped from 4,800 to 400 in the De Wendel plants between 1946 and 1950.[36] Thus, the counterpower that could have prevented the return of paternalism disintegrated until the late 1950s.

The organization of social life in the steel towns revolved around the plant and the *patron*. Capital was concentrated in small uncompetitive plants with labor-intensive technologies, and owners sought a political solution (market sharing) to competitive threats. Labor practices were governed by a search for social stability. Employers tended to own essential elements of the social infrastructure. Housing, central to the paternalist system, was provided to most workers. According to one estimate, the steel firms throughout France could house over 50 percent of its employees by the mid-1950s.[37] The figure was higher in Lorraine, especially if one includes the numbers of immigrant workers living in dormitories. Most of the units were segregated so that workers, clericals, technicians, and engineers all lived in separate areas. Housing was contingent on the labor contract. Loss of job threatened an entire family with eviction—a powerful incentive to compliance.

Employers had other weapons in their arsenal. Companies owned stores, ran schools, and even paid the clergy. Since many churches depended on the benevolence of the "*seigneur*," the Catholic hierarchy, especially older priests, preached a sermon of duty and responsibility. Not even agnosticism provided an escape. Given the special status of schools in the Moselle (due to the German occupation), the Church had a major impact on public education as late as the mid-1970s. Religious education was required.[38] Local firms also subsidized middle and secondary schools. Employer associations ran the vocational training centers through which young workers passed to gain apprenticeships. Countering the moral authority of the *patron* required social resources that were not easily available. Eugène Descamps, who headed the Lorraine steel workers for the CFTC and who later became secretary general of the CFTC/CFDT, described his reactions to the paternalist system when he arrived from the Nord to Lorraine in 1950:

> One thing offended me upon arriving: the weight of the "De Wendels" on people. In cities such as Hayange, Moyeuvre-Joeuf, everything is De Wendel. One is born at the De Wendel hospital or at the De Wendel clinic, one goes to the De Wendel school, one is apprenticed at the De Wendel training center, one becomes a worker at De Wendel, and when one dies one is entitled to a De Wendel wreath. The hold of the plant on men, on the young, on everything was unbearable for me.[39]

This hold was reinforced through a number of employer sanctions. Breaking the employment contract entailed local blacklisting. It brought the loss of housing, potentially devastating until the housing stock was increased through public assistance. It threatened the job possibilities of male children, since the apprenticeship system promised continued family employment. Workers without jobs had to leave the area. In the steel towns, few industrial jobs were not in some way connected to the plant. Because public transportation in the 1950s was also poorly developed, the companies tended to provide bus transport. Rail links between towns were used almost exclusively for freight.

Paternalistic employers sought to prevent the growth of alternate sources of legitimacy. Unions were the real enemy. Union membership put workers outside of the "family." It also threatened the employment relation. Antiunionism carried down through all levels of supervisors. René Carême, a longtime CFTC and later CFDT organizer, tells the story of what happened when a foreman discovered that he had attended a collective bargaining session: he was told "never to stick his unionized nose" in the shop again.[40]

While Carême was kicked out of only one shop by an overzealous foreman, other militants were not so fortunate. Although postwar labor legislation allegedly protected members of works councils or *délégués du personnel*, the steel companies were not prepared to adhere to legal niceties. De Wendel fired fifty-two organizers after the 1950 strikes. The Sidélor plant in Micheville pursued several militants in court after a seven-week strike in 1951 and secured imprisonment.[41] The harassment discouraged union members from running in firm-level elections. After the De Wendel firings, for instance, the CGT could get only 20 percent of the candidates it had the year before to run for *délégué* positions.[42] Workers feared the consequences of union activities.

The employers tried to avoid such dirty work by finding compliant workers in the first place. When they recruited workers from other towns in Lorraine or Nord, from other parts of France, or from abroad, they carefully checked backgrounds. The unions were particularly upset by how employees for the new Sollac plant were chosen. They claimed an attempt was made to find the most docile workers from Brittany and Algeria.[43] On several occasions in the early 1950s, employers secured the deportation of immigrants for unruly behavior. The labor shortages after both world wars forced employers to recruit actively outside the local industrial working class. The most immediate group was the Lorraine peasantry. A high number of workers in the early 1950s were worker-peasants who retained their self-reliance, their suspicion of unions, and a view of industrial employment as temporary.

In Lorraine, the number of immigrants as a percentage of total population was the highest for any region in Europe by 1955.[44] Waves of Italians came to work in the mills in the 1890s and 1920s, with a smaller influx continuing after 1945. In the Nord, Poles arrived in the 1920s as well.[45] North Africans (mostly from Algeria and Tunisia) were imported after the 1920s, with growing numbers arriving in the 1950s. Unlike the Italians, who moved quickly into skilled jobs in the plants and who developed tight-knit subcommunities outside, colonial immigrants filled mostly unskilled jobs and lived a miserable existence in male dormitories.[46] Language and cultural differences impeded easy integration into local working-class culture. The Belgians, who crossed the border daily, especially for work in Longwy and Denain, constituted a final group of immigrant workers. Unlike other ethnic groups, these workers began in skilled employment. Outside the fabric of local communities, the Belgians had little interest in jeopardizing their employment contract. The percentage of foreign steel workers in France remained the highest in the ECSC until the 1970s.

Labor shortages created a tension between the social power of employers and the market power of scarce labor. The paternalist system tied the most valued workers to the firm and formalized a dual internal labor market. The scarcity of labor therefore benefited only skilled workers and certain categories of semiskilled laborers. The North Africans, 85 percent of whom were unskilled and half of whom were illiterate, suffered considerable unemployment during recessions.[47]

Employers were not entirely successful in imposing social quiescence. Strikes took place, although they varied in frequency. The major national strikes of 1947, 1948, and 1950 were as bitter in the steel mills as elsewhere in France. Strikes over local issues were more consequential to the development of the steel working class. There is little hard data on strike volume because work stoppages tended to be short, involved small groups of workers, and were not recorded in a systematic fashion by labor inspectors. We must rely, therefore, on impressions. Interviews with militants active during this period indicate that few plants experienced social calm.

Thus, paternalism made for struggles. The *patrons* never succeeded in completely eradicating local or imported class militancy. Employers could not always use ethnic outsiders to divide workers. While immigration rates were high, they were uneven. Even in the Longwy basin, for instance, the Rehon plant had less than 5 percent foreign workers, while five miles away 50 percent of the work force at the Mont St. Martin plant were immigrants. Moreover, the tangible benefits of paternalism were not always evident: material conditions for the industrial working class deteriorated in the late 1940s and early 1950s. Purchasing power did not stop falling until 1955.[48] The Korean War boom increased work cadences so that workers' purchasing power increased because of overtime.

Heavy-handed intimidation by employers contributed to the formation of class relations in Lorraine. The paternalist system helped foster the class anger of the CGT and legitimated the class-based opposition of the PCF. It also nurtured a sense of moral outrage in the CFTC, which might otherwise have followed the strategy of fellowship pursued by CFTC unions elsewhere. It contributed to violent tactics and bitter negotiations and ensured state intervention to maintain social calm.

Steel paternalism was finally put to rest in the 1960s by five specific factors. First, the diffusion of state services—especially low-cost public housing and medical care—weakened employers' control over key services. Second, the new managers hired from the late 1950s onward were increasingly graduates of the *grandes écoles* and consequently found the traditional social policies distasteful. Third, the development of mass media in France introduced other models of behavior for workers who might have been tempted by the employment policies. Fourth, state intervention to restore the industry to solvency (see chapter 5) forced the firms to sell off housing. By the 1970s, the steel firms were disengaged from most of their workers' outside activities. Fifth, material conditions improved after the mid-1950s, and real wages rose during the industry's brief decade of prosperity. Prosperity made workers and their families all the more difficult to control.

Steel employers sought to maintain a large core of skilled workers because their skill was embedded in experience. Workplace authoritarianism also dovetailed with "divine-right" approaches to management in the family firm. Paternalism died for a variety of reasons, most of which were associated with modernization.

Conclusion

In the twenty years after 1945, French steel companies were exposed to European competition, yet they managed to carve out their own national market. They also guarded successfully their managerial autonomy against a potentially encroaching state and a class-conscious labor movement. They seemed to defy the tools of the strong state and defeat the passion of the modernizers. Yet economic growth masked a competitive weakness that would eventually undermine the edifice, as we will see in following chapters.

The weight of historical choices was particularly burdensome for French steel. The very success of self-governance later impeded industrial development. Small plants, older technology, and a work force that was both overskilled and underskilled slowed the growth of optimal production units. Production for national markets assumed a steady increase in domestic demand, stable trade patterns, and a continuing comparative advantage. These assumptions eventually proved to be short-lived.

The mismatch of policy tools undermined the state's attempts to modernize steel. Employers' labor policies impeded the rationalization of work. The combination of industry self-government and working-class exclusion might be an explosive mix under different social conditions—a nonexpanding steel market, for instance. Both labor and business, however, believed that the state held the keys to industrial change.

FOUR

Labor's Response to Exclusion

The political bargains struck over industrial adjustment in the early postwar steel industry preserved managerial autonomy in both product decisions and labor policy. At a time when the government was unwilling to get involved in workplace relations, a weak and internally divided labor movement had to find ways to influence the conditions of employment on its own.

At the height of labor exclusion in the 1950s, working-class mobilization involved strategic dilemmas. Union organizers had to balance the benefits of a political alliance between radicalized workers and leftist political parties against the need to increase cooperation between unions and various political orientations. They faced a related tension between bread-and-butter unionism and class-based militancy: the unions criticized owners for lackluster performance, but union opposition encouraged militancy that discouraged modernization.

Failure to modernize and implement mass production perpetuated paternalism. A combination of repressive labor policy and the dispersion of union power diminished the effectiveness of both class-based militancy and bread-and-butter unionism. Owners managed to keep union organizations at a distance, delay the coordination of collective bargaining, and maintain control over labor policy until the mid-1960s. Yet labor leaders learned how to work around union weaknesses to preserve their membership and to increase their bargaining leverage. Militants from the CGT constituted a partial counterweight to employer paternalism. Workers joined forces for collective action to bring employers to the negotiating table. They learned to use divisions in the labor movement to force individual employers to bargain more responsibly.

Labor benefited from industrywide growth and consolidation from the mid-1950s to the mid-1960s. The new generation of managers was less paternalistic and more interested in modern labor management techniques. Rising profits, a strong demand for skilled workers, and growing faith in the value of mass consumption increased union success at the bargaining table. Consolidation of assets led to de facto centralization of negotiations. Just as the steel industry rose from obsolescence to prominence between 1945 and 1960, so steelworkers journeyed from disarray and exclusion to partial participation in the golden age of steel. But the boom disguised fundamental flaws in labor relations, just as it concealed competitive weaknesses in steel production. When times turned tough in the mid-1960s, labor turned to the state to compensate for internal contradictions.

Labor and Communism in Steel

The paternalist system depended upon controlling the organizational and political power of the workers. The avowed enemy were the unions, which constituted a counterauthority resisting the employers' domination. Given the state's indifference and the industrialists' influence over most other social institutions, labor had few allies in the struggle to improve working conditions. Thus, unions had both practical and ideological incentives to join forces with leftist political parties in challenging property rights and restructuring social relations. The strength of union-party alliances varied according to the demographic and political characteristics of the steel regions. Relying on Left political parties to represent labor's interest spurred oppositionalism without facilitating collective discipline, interunion cooperation, or effective action.

The nexus of the CGT and the PCF challenged the employers in the steel regions. The Communist Party constituted an important source of support. Both union and party shared a vision of capitalist society that provided a concrete recipe for local activism—opposition to the initiative of the steelmakers. Not only did both criticize the abuse of property, they constantly cast aspersions on the patriotic credentials of the steel managers. Yet opposition was always tempered by a pragmatism that was necessary to sustain support.

Aside from its ideological correspondence with the largest union confederation, the Communist Party played a practical role in industrial relations in steel: its municipal councils supported industrial actions. Support could take many forms, ranging from free lunches in a soup kitchen to summer camp for children of strikers. While Communist mayors and city councils usually avoided overt favoritism toward particular local interests, they used their resources to support groups that shared their political agendas. Commenting on municipal politics in general, one Communist mayor contended:

> There are . . . the facilities that a communist municipality offers willingly to democratic organizations: offices, headquarters, organization of congresses for local, regional, even national groups. These are the important advantages that permit different democratic political formations to develop their activities in the best of conditions. . . . The primary interest [in running local government] resides in the possibility . . . of solidifying our policy in the service of the interests of the population. . . . In particular, the social character of our policy can find an important expression.[1]

By "democratic organizations," of course, the mayor meant groups on the left. Depending on the political climate, the term might include other unions as well as the CGT. Later elected Communist officials were active in publicizing the plight of workers in regions dominated by one industry and in lobbying prefects after the mid-1960s.

At least in Lorraine, the Communists' role in working-class politics was different from the national pattern: the "transmission belt" ran from union to party. The party in Lorraine derived its strength from the CGT.[2] When the CGT picked up votes in workplace elections, it was followed by a Communist advance in local elections. Workers typically joined the union before committing to the party. Consequently, party members had a good understanding of shop floor issues.

Naturalized and second-generation Italians contributed more than any other subpopulation to the vitality of the CGT and the Communist Party in the Lorraine iron and steel region.[3] Serge Bonnet attributes this anomaly to the radical origins of Italian immigrants, many of whom left Italy for political reasons. They did not emigrate from the Mezzogiorno but instead came from the industrial regions of the north of Italy, carrying with them a historical memory of industrial conflict.[4] The union and later the party became the vehicles through which first- and second-generation Italians were integrated into French political life.

Consequently, both party and union had reliable electorates from within the Italian community. The Italian "core" in areas such as Longwy created a working-class movement based on personalistic norms as well as shared material interests. In contrast, union and party had more difficulty organizing other immigrant groups, in particular the North Africans.

Communism was strongest in the smaller company towns of Lorraine. Even Longwy, which became known as "the people's republic," had a lower percentage of Communist voters than did Villerupt or Briey. Radicalism was also more prevalent among iron miners than steelworkers. Part of the difference can be accounted for by the higher percentage of Italians in the mines. Mining also requires closer, more cooperative working teams. Finally, Communist support was highest among the skilled and semiskilled workers, which partially accounts for support by the Italians and indifference among North Africans working in lower-skilled jobs.

Support for Communist candidates in local elections was class-based. The change from conservative Catholicism to communism as a force in electoral politics occurred in less than a generation—between the late 1920s and the late 1940s. The dramatic shift at least partially reflected the shallow roots of French socialism in Lorraine. As Bonnet shows, rising support for the Communist Party correlated strongly with increased immigration.[5] The CGT facilitated the growing strength of the PCF in municipal councils through its plant-level support. Some towns, such as Hagondange, had been led by the Communists since the war. Others, such as Moyeuvre (considered a De Wendel town) were captured in the late 1950s. The PCF captured Villerupt in 1959.

In the Nord, by contrast, the regional Socialist Party had a reputation for its municipal machines, against which the PCF had long struggled. Communists after 1945 gained support in formerly socialist districts, while the socialists maintained some of their strongholds. Thus, both the PCF and the CGT were better implanted in the Nord. The dominance of the PCF and CGT in the Nord steel towns made compromise unnecessary. The CFTC/CFDT worked more comfortably with the FO because of its distinctly minority position. These traditions were actively reflected in the behavior of union militants.

Because party strength derived from workplace position, the platform of the Lorraine PCF has always been less strident, less dogmatic, and more concrete than that of the national party, although differences are frequently obscured by intraparty politics. Most important, however, the Lorraine party did not try to control the CGT. Therefore, to the occasional embarrassment of local Communist leaders, the class discipline that the PCF ordinarily could enforce was lacking in the Lorraine steel regions. The party could not quell the direct actions that erupted in the late 1970s (see chapter 6).

Because the PCF in Lorraine was not the linchpin of working-class politics, industrial relations assumed a different hue from that found elsewhere in France. First, the CGT was not as willing in Lorraine to transform shop floor struggles into national political issues. The CGT stayed closer to industrial issues in the Lorraine steel industry than it did in other sectors. The party (not the union) pursued the national political issues. Partly because its leaders came from the base, however, even the Lorraine party tended to stay close to local issues.[6] Thus, strikes in Lorraine tended to be less frequently "mass action" or "days of action" organized from Paris than elsewhere in France. Instead, they were more related to shop floor and industry demands. Second, CGT relations with the CFTC were more amicable in Lorraine than in the Nord, for instance, where a tradition of enmity developed. Better interunion relations favored the development of joint action in Lorraine.[7]

Both the CGT and the PCF fought employers' attempts to impose social control after the war. Yet the system of industrial relations did not facilitate collective discipline within the labor movement. The struggle between paternalism and oppositionalism had lingering effects well after the material conditions for either strategy had been eliminated.

Competition and Cooperation among Steel Unions

The uneasy alliance between radical workers and Left political parties exacerbated the rivalry that characterizes interunion relations in France. The three major industrial unions disagreed over both programmatic and practical issues. Given the oppositional tactics adopted from Left political parties, the constant politicking among the unions generated shop floor agitation and facilitated mobilization. However, labor had trouble translating agitation and social mobilization into programmatic goals and bargaining leverage. Early efforts at interunion cooperation, both in day-to-day tasks and in several major strikes, highlighted the differences dividing the CGT, the CFTC, and the FO. At the same time, these experiences convinced union leaders of the need for some degree of cooperation if workers were to influence the conditions of their employment.

The first postwar decade was characterized by sharp political divisions. The most traumatic element in this period for the labor movement was the rehabilitation of the business community. The "steel barons" reestablished social control and regained their previous political influence. The French state seemed prepared to support the owners against their workers.[8] Both the CGT and the CFTC refused to meet with industry representatives in the modernization committees of the Plan between 1951 and 1954. The union boycott was a partial response to the industry's refusal to engage in collective bargaining with the unions during this period. Layoffs of union militants by De Wendel after the local industrial action of 1950 aggravated the tension.

The CGT enjoyed a dominant position within the labor movement until the 1960s. The CGT was not afraid to articulate concrete demands on the shop floor. It was also the only union to stand up to employer paternalism before the 1950s. The CGT built its support after the war, as it did elsewhere, on its record of wartime resistance. Until 1948 its only rival was a CFTC which, until 1950, acted more like a Christian brotherhood than a trade union. Even the split in the CGT that created the FO barely diminished its support.

Before the Second World War, the CFTC had hardly existed in the steel regions of Lorraine and the Nord. Its only bastion of support was in the coal mines of eastern Moselle. However, the Young Christian Workers (Jeunesse Ouvrière Chrétienne, JOC) had been active in the Resistance, especially in Lorraine. The Lorraine JOC was among the strongest in France by the late 1940s. Out of its ranks the CFTC pulled Eugène Descamps to head the Lorraine steelworkers. The future secretary general of the CFTC/CFDT and other progressive elements of the Reconstruction wing of the CFTC molded an organization in the steel regions much closer to CGT unionism than the prewar CFTC "brotherhoods," or even the company unions of the coal workers.[9]

The CFTC steel unions attracted support by aggressively pursuing bread-and-butter issues. Unlike the CGT, its strike calls came only over salaries and working conditions.[10] Unlike the FO, however, the CFTC rarely criticized strikes or condoned employers' discipline of CGT organizers. The CFTC's actions were militant and based on embryonic theories of class struggle. In contrast to the CGT, the CFTC metalworkers supported the ECSC.[11]

The CFTC faced considerable employer resistance to its organizing efforts. Its initial penetration was in Catholic Moselle, an area in which the CGT was traditionally weak. Departmental archives show the CFTC to have been particularly combative at the Hayange plant of De Wendel, at Lorraine-Escaut's Thionville plant, and at Sollac.[12] Descamps convinced the traditionally white-collar elements in the CFTC to recruit actively among the North Africans. Joint action by local unions brought a larger potential audience for the small organization. The development of CFTC/CFDT organization and support took a generation. Already by the early 1960s, it had captured works councils in the Moselle. However, because of local peculiarities, the influence of the CFTC/CFDT spread unevenly. In areas such as Longwy, the CFDT did not have a breakthrough of support for another twenty years.

The FO remained the ghettoized union after the scission of 1948. Although declared a representative union nationally which qualified it to bargain regionally, the FO rarely gained more than 5–7 percent in professional elections until the 1970s. In a number of plants, especially in the Meurthe-et-Moselle, no plant organization at all existed. When it broke off from the CGT, the FO had brought perhaps 2,000 members.[13] Many of these eventually returned to the CGT, partially because of the blind anticommunism of FO leaders, a position that threatened the entire social fabric of working-class culture in Lorraine. René Schwob led the FO steelworkers in Lorraine until the mid-1960s. Before the FO left the CGT to form its own organization, Schwob had been widely respected within the CGT for his mastery of technical details and his capacities as a negotiator. After the split, Schwob acted as much on his own virulent anticommunism as on his negotiating ability. Joint action rarely took place with the CGT. While FO members infrequently crossed picket lines, its works councilors frequently sided with management in disciplinary actions against CGT militants within works councils. The FO continually denounced the close relations between local Communist governments and CGT ULs.

The 1947 and 1948 strikes tested interunion relations. They resulted partially from the national mobilization effort by the PCF and CGT to protest the expulsion of the Communists from government and the course of reconstruction. They also reflected anger within the workplace at the workers' deteriorating material conditions in an industry deemed critical to reconstruction. Both the CGT and PCF had intervened in local disputes in 1946 and early 1947 to defuse protest and maintain stability, and the subsequent strikes were largely a reaction.

The December 1947 strikes were generally considered a failure. Both the CGT and PCF pushed the strike actions to protest Marshall Plan aid, the "right turn" of the governing coalition (without Communist ministers), and the cold war. The strike action was coordinated from Paris, in all likelihood by the Confederation.[14] It was part of a national-level mobilization that started in the public sector, moved to the docks, then to the Parisian metalworkers, and finally to the coal miners. Although consultation with the rank and file was lacking at certain metalworking plants, militants in the steel plants gained a consensus backing a walkout. Wage demands were issued, but the strike actions were clearly developed outside the steel regions. The government used both army and police to break the picket lines and imprison party and union militants. The strikes and the behavior of CGT and PCF militants forced the secession of the Force Ouvrière by mid-December.[15] While the strike action in steel was not as bitter as in other industrial sectors and regions of France, it nonetheless set an important "lesson" for future behavior. The strikes came to be identified with the PCF, and they failed because of state repression.

In contrast, the strikes of September and October 1948 were coordinated by the CGT, the CFTC, and the Mouvement Populaire des Familles, a Catholic social organization. The state responded to the strike activity by sending police into the plants to assume security precautions. The refusal by employers to negotiate wage increases provoked workers in one coking plant in Micheville to let the facilities deteriorate in an act of sabotage.[16] Class violence had outstripped the restraining capacity of the unions. In the Longwy region, the wives of 200 militants took several managers hostage, and "convinced" them to seek mediation from the prefect. The next day, the employers acquiesced to salary increases in collective bargaining session.[17] Neither the CGT nor the CFTC condemned the hostage taking, although the CFTC was uneasy about industrial sabotage. The FO opposed both forms of protest.

Labor unity was never easy. The CGT had long seen itself as the sole representative of the working class (in steel and elsewhere). It tried first to ignore the breakaway FO. When the latter began signing contracts, the CGT dismissed it as traitorous. The CGT had fewer suspicions about the CFTC, but still tended to argue in favor of a "*syndicat unique*" that it would control. There had been sporadic cooperation during the 1947 and 1948 national strikes as well as the 1950 De Wendel strike despite political differences and organizational rivalries. Locals from both the CGT and CFTC cooperated in twenty-four- and forty-eight-hour strikes to support the public service strike called at the confederal level in August 1953.

By 1954 the CGT and CFTC were planning joint bargaining positions over minimum salaries, wage increases, family allocations, and abolition of regional inequalities.[18] For the CGT, unity legitimated its own claims of class universalism. The CFTC needed unity to expand its audience. Thus, trade-union unity—at least between the CGT and CFTC—was a necessary complement to a greater focus on concrete demands. The effort to maintain unity still remained a tense struggle until the late 1960s.

Mobilizing for Growth

The disappointing results of the first postwar decade caused the steel unions to rethink their strategies for combating workers' exclusion. Union leaders needed to maximize their scant resources in the struggle against paternalism, but a fundamental tension existed between efforts to ally with Left political parties and to increase labor unity. The CGT and CFTC could cooperate most effectively when they focused on bread-and-butter issues and minimized the class-based militancy that came from the CGT-PCF connection. Yet a piecemeal approach to improve wages and other working conditions would not challenge the paternalistic exclusion that prevented effective negotiation over the concrete labor issues on which the unions increasingly focused.

In the excitement of the Liberation, the CGT had cooperated in early attempts to revitalize the industry and boost productivity. Many militants had sat on the purge committees during the Liberation to evaluate the collaboration of the employers who tried (or were forced) to maintain production during the war. In the plants, the task for the CGT was to restart production to begin the process of reconstruction. This coincided with the confederal support for the national planning process. The CGT found itself in the position of dampening militancy over the course of reconstruction.

The rewards for such a strategy were meager. By 1947 the industry was seeking to hold the line on wage increases. However, the Resistance coalition had broken apart in 1946–1947, and the Communists were expelled from the coalition government in May 1947. Exclusion of the radical Left removed organizational constraints on militancy. The strikes of 1947 and 1948 forced some wage concessions from employers. Strike activity until the early 1950s sought wage increases to keep pace with inflation.

The CGT Metalworkers (Fédération des Travailleurs de la Métallurgie, FTM) rarely missed an opportunity during this period to call for nationalization of the industry and to remind the political community that nationalization had been part of the Resistance program.[19] Such calls were made at first on patriotic grounds.[20] Later, arguments concerning national loyalty were transformed into arguments based on efficiency. According to the CGT, an obsession with profits led the industry to neglect its competitive capacity.

The CFTC Metalworkers (Fédération Générale de la Métallurgie, FGM) also argued from this early stage for the nationalization of steel. Such a position was a bit more tenuous for the FGM, because its confederation opposed nationalization on principle. Yet, as early as 1952, the FGM argued vehemently against the "holy employers of the steel industry."[21] The CFTC-FGM was as appalled by the reimposition of the paternalist system as by the inefficiencies of production.

The theme of nationalization was pushed aside by both the rapid changes affecting the industry and the unlikelihood of its realization. For the CGT, the creation of the European Coal and Steel Community constituted an unmitigated disaster for French industry and the French nation, which would place the steel companies under the domination of the Krupps and Thyssens.[22] The CGT predicted that German products would overrun the French market, a position quite similar to that of the CSSF. In line with the positions of the PCF, the CGT also argued that the ECSC meant rearmament, especially given the Korean War and later the proposals for a European Defense Community. The ECSC implied an aggressive stance against the Soviet Union and a threat to international peace.

Such explicitly political positions sat poorly with French steelworkers, regardless of their political sympathies. Delegates to the Eighteenth CGT-FTM Congress in 1952 argued that the union's positions—either from the Confederation or the Federation—were not particularly effective in addressing shop floor demands.[23] The CGT-FTM sought instead to make a connection between the job losses that resulted from the concentration of production in the early 1950s and creation of the ECSC.

According to the FTM, the Treaty of Paris would result in the "deindustrialization of France."[24] The union had proof: the small specialized steel plants of central France suffered considerable unemployment in the early 1950s. This was a touchy issue, however, for the Federation. Employment in the industry as a whole was growing—reaching a peak in the mid-1960s—with the gains made in Lorraine and the Nord. Thus, the CGT focused on the displacement of workers and not on increased employment elsewhere.

In contrast, the CFTC favored the ECSC on efficiency grounds. The expanded market promised to provide access to German coal and coke, forcing the relatively inefficient French steel producers either to lower prices or produce more efficiently. Such lower prices would help steel users.[25] Therefore, against the owners' trepidations, it argued for an extension of the market. Against the CGT/PCF positions, it pointed to the social guarantees of the ECSC.[26]

The FO derided these political positions. René Schwob ridiculed CGT opposition to the ECSC as mimicking the PCF's position.[27] The FO welcomed the larger European market for its efficiencies and scoffed at threats to national sovereignty. To distance itself even more from the CGT, it publicly thanked the Americans for Marshall Plan monies.[28] The FO would instead throw its energies into collective bargaining and seek higher wages for steelworkers.

In the wake of failed struggles over nationalization and the ECSC, the CGT sought more concrete labor market strategies. Along with labor displacement in the older shops, higher wages and greater purchasing power were the most significant demands of the CGT-FTM by the mid-1950s. After the early 1950s, the CGT pursued "small demands next to general demands."[29] The argument for a new focus on collective bargaining had two parts. First, political strategies had led to a demoralization of militants and general social demobilization. Second, the owners by the mid-1950s were recording profits and paying out dividends. They could afford higher levels of remuneration.

Nonetheless, by the mid-1950s union leaders recognized that their organizations needed to focus on more concrete issues. Defeat in political campaigns led to social demobilization. The CGT had lost membership consistently between 1947 and 1953.[30] Strike calls over imprisoned leaders met with disapproval from workers worried about feeding their families. Political positions bordering on sectarianism hurt the FTM in steel as much as it did the Confederation.[31] Moreover, by the mid-

1950s, production was increasing annually, and the industry had the outer trappings of prosperity.

Still, because institutional memory is such a powerful force, organizations find it very difficult to reverse previous strategic positions. The newfound focus on wages and collective bargaining did not carry with it a depoliticization of industrial relations. On the contrary, both the CGT and the CFTC continued to argue for nationalization. The CGT remained opposed to the ECSC and the German threat to French sovereignty. The memory of the struggles between 1947 and 1953 took on legendary proportions in later years. In the Longwy-Villerupt area, according to Gérard Noiriel, these strikes affected the militant consciousness even more than the Popular Front.[32]

The brief decade of relative prosperity for the industry refocused the general orientations of the CGT and CFTC. In the mid-1950s, the CGT claimed that the upturn was only temporary, warning workers of the dangers of the ECSC and German penetration of the French market.[33] It later refined its analyses to focus more on the choices of steelmakers rather than imminent dangers. It contested, for instance, the decision to choose coastal "greenfield" locations, away from existing plants and equipment, for new industrial sites. Dunkerque in the late 1950s and Fos in the mid-1960s became threats to the livelihood of Lorraine. The CGT argued that steelmakers needed to reinvest in the older communities and that the *maîtres de forge* would deindustrialize Lorraine as they had depleted central France in the late 1940s. In contrast, the CFTC took a different tack, arguing (not too loudly in deference to its Lorraine audience) that such investments were necessary for the industry to remain competitive.

Both the CGT and the CFTC mimicked the productivist rhetoric articulated by industry managers. Like the annual reports of the companies, union publications consistently glowed over record-breaking tonnages. However, both the CGT and the CFTC argued that the companies should not pay dividends and should instead reinvest earnings to increase capacity. The industry was held responsible for its slow growth relative to Germany.

Both unions saw profit levels as excessive and collective bargaining as whittling away at that surplus. The CGT, convinced that the owners could pay virtually any figure, formulated "hard" positions. (This will be discussed later.) The CFTC took the moral perspective: industrialists had an obligation to their workers, having robbed them of dignity through paternalistic practices. Both unions were convinced that the owners refused to bargain in good faith.

In contrast, the FO was more willing to explore the industrial choices facing the industry. The terms of competition, according to the union, had changed by the early 1960s. The industry needed to move into higher value-added products because producers from the less-developed countries were capturing the low-end market for crude steel. In the meantime, price controls constricted the companies' ability to finance investments.[34] The FO scoffed at the CGT's and CFTC's insistence that the industry was making undue profits. The industry's position needed to be disaggregated; some firms were doing better than others. If those in a weaker competitive position could not meet their financial obligations, they would go bankrupt. Industry needed (1) a rollback of price controls; (2) assistance in importing cheaper raw materials; (3) resistance to the protectionist temptation to preserve foreign markets for French products. The FO's analysis borrowed heavily from the industry's own assessment, since the FO was the only union that took the planning sessions seriously, and it met regularly with the employers to hear their investment plans.

The emphasis on bread-and-butter demands did not eliminate national politics from the union movement in the steel industry. Unlike the tumultuous years after the war, however, in the late 1950s and early 1960s politics did not directly affect the industry and the welfare of its workers. Both CFTC and CGT locals actively opposed the Algerian War, organizing demonstrations and showing solidarity with "democratic forces." Both had serious reservations about de Gaulle's assumption of power, with the CGT moving into opposition in 1958, especially after the beating the PCF took in elections of that year. Mobilization against the "ultras" in the Algerian conflict and the inauguration of a new regime further radicalized the Lorraine CFTC-FGM, the core of the Reconstruction wing of the confederation that would introduce democratic socialism to Catholic trade unionism.[35]

The articulation of overtly political positions guaranteed dissension within the French labor movement. This was true in the CGT's sectarian phase between 1947 and 1952 and in the early de Gaulle years. The leaders of the CGT and CFTC, Benoît Frachon and Eugène Descamps, respectively, carried on a running feud over links between the CGT and PCF.[36] Relations between federations were only a little smoother during the miners' strikes in 1962–1963 (see chapter 5). Local CGT-PCF coalitions rankled the CFTC, which objected both to the PCF's platform and the material aid given by the Communists to the union.

Working-class politics suffered from a cruel dilemma in the Lorraine steel industry. Unions needed a solid front to confront paternalistic labor practices, and the CGT-PCF nexus provided the core of such opposition. Yet the stronger that nexus, the more difficult became interunion relations. To the extent that the CFTC began to represent steel workers, such relations became a gauge of working-class unity. Those relations, in turn, functioned smoothest when the issues remained out of the political sphere and within the labor market. The only way to bargain over concrete issues, however, was to dislodge paternalism, and that required an appeal to a higher political authority. As the unions mobilized for growth in the steel industry, they also mobilized for change.

Early Forays into Collective Bargaining

The first attempts to bargain collectively stumbled on employer prerogative and acrimony among unions. The steelmakers used regional bargaining and plant-level wage setting to shape labor policy in ways that preserved profits and protected noncompetitive plants. When labor negotiations were unavoidable, they tried to strike deals with the FO, the most compliant of the unions. Yet labor learned to use oppositionalism and interunion competition in effective bargaining tactics. For a brief period, circumstances enabled a weak labor movement to secure some bread-and-butter benefits on its own. But early union successes still depended on industry growth and management's tolerance, two conditions that would not last for long.

Collective bargaining in the early postwar period responded more to social mobilization than to a need to negotiate wage levels. Both unions and employers paid lip service to the virtues of collective bargaining. The employers were not yet fully rehabilitated from the wartime experience, and the unions wanted to act "responsibly"—at least while the Communists remained in the government.

Both unions and employers, however, relied more on confrontation than on regular negotiations. The strikes of 1947 and 1948, while reflecting the material hardship of French workers, also tested the magnitude of trade union power. The owners sought to dampen wages because other raw material prices and finished product prices were both set administratively. Moreover, until the 1950 framework law, collective bargaining was not "autonomous": all agreements needed government approval.[37] A year after the law's enactment, the steel employers withdrew from bargaining sessions because of dissatisfaction with the government's handling of minimum wages (as called for in the law). Minimum wages had been set higher than inflation, thereby placing pressure on contract wages. Between 1951 and 1954, no meetings took place while the owners assessed their positions vis-à-vis the law and the unions. For three years, owners set wages unilaterally.

The nature of collective bargaining sessions during the decade of prosperity (1955–1965) conformed to the dominant national pattern of bargaining.[38] There was a division of labor among employers' groups. The CSSF occupied itself with investment, finances, and production while it left collective bargaining to six regional Chambres Syndicales, which belonged to the peak metalworking employers' federation, the Union des Industries Métallurgiques et Minières (UIMM). However, the same individuals—owners or board members—overlapped in their jurisdictions. This division of responsibilities enabled the employers to "regionalize" bargaining. The policy was consistent with those of other industries in the UIMM.[39]

According to François Sellier, French metalworking employers have long sought to avoid both sectoral- and plant-level negotiations, seeking instead regional and national deals.[40] Sectoral discussions were dangerous because they permitted unions to bargain on behalf of all workers in an industry, set minimum standards of competition, and remove wages from competition. The UIMM avoided plant-level accords because they tended to pit employers against each other.[41] The UIMM preferred discussions at the regional and national level. Regional discussions permitted the stabilization of local labor markets and protection of marginal firms. Lorwin noticed this tendency in the early 1950s:

> Area bargaining fortifies the tendency to align the contract wage scales (or unilateral recommendations) of employer associations on the wages of the marginal firm; . . . in general association wage rates are set in line with the ability to pay, real or fancied, of the marginal enterprise. Even when there are departures from such minima, association rates give the tone to actual wage rate movements.[42]

Because of loose language, national framework accords covering all metalworking industries protected the weakest firms and sectors. Regional bargaining units did not guarantee successful negotiations. Only the *chambres syndicales* in the Nord and Lorraine signed agreements with the unions. Workers in steel plants in central France were covered by departmental collective bargaining agreements for the metal trades (that is, between the UIMM and the metalworkers' federations).

Although the steelmakers disliked plant-level bargaining, they actually sought plant-level wage setting in which they could establish (not negotiate) real salaries. The starting point for wage negotiations, both nationally and in the steel industry, were *minima*—minimum wage levels for each broad skill category, of which there were usually six to eight. Attached to these *minima* were various bonuses granted according to seniority, shift (day, evening, or night), and the degree of difficulty or danger associated with work. While some bonuses were negotiated through collective bargaining, others were set unilaterally at the plant.

French collective bargaining distinguishes between a collective agreement (*convention collective*) and a wage settlement (*accord de salaire*). According to the 1950 law, the former must include provisions on union rights, working conditions, equal pay, conditions of employment, training, vacations, etcetera. The *accord de salaire* is not obligatory. The cumbersome requirements of collective agreements encouraged quick wage settlements.

The resumption of collective bargaining in 1954 led quickly to a joint agreement for the East. The Chambre Syndicale de la Sidérurgie de Moselle and the Chambre Syndicale de la Sidérurgie de Meurthe-et-Moselle signed a *convention collective* with the three unions on which all other agreements were based. The two *chambres syndicales* had every reason to work together, for the labor markets were quite comparable in the two departments. After 1954, bargaining took place with departmental *chambres syndicales*, although they regularly coordinated activities and aligned wages. The two departmental employer groups of Lorraine merged in 1962 to form the Association de la Sidérurgie et des Mines de Fer Lorraine (Assimilor), an organization that also combined the employer groups for the iron mines in the two departments.

Negotiations took place regularly in the years after the *convention collective*. These typically focused on *minima*, although agreements on bonuses, overtime, and vacation pay were also signed. The results are difficult to interpret. Real wages in the steel industry did not seem to rise much faster than the national average. Still, real salaries doubled between 1954 and 1962 and would double again by 1971.[43] Moreover, workers received three and a half weeks vacation by 1956 and over four weeks by 1963. Likewise, the work week was reduced from fifty-six to forty-eight hours in 1957, although the average work week included considerable overtime. Both vacation increases and worktime reductions preceded national-level accords that generalized these benefits to the entire work force.

The FO did most of the successful negotiating until the 1960s. The CGT attended meetings but signed few of these wage or "wage-plus" agreements. The CFTC played both strategies, signing numerous agreements (although fewer than the FO) while frequently siding with the CGT in its "hard" positions. The CGT felt it had more to gain by withholding its signature. It could consequently stake out a class position. Its opposition, however, also benefited the signatory union, since the latter could always point out to the employers that an aggressive CGT wanted even more than it did.

The oppositional unionism of the CGT created a set of negotiating symbols that pressured the negotiating parties. Employers knew that an insufficient agreement might lead to industrial actions led by the CGT. The FO knew that dissatisfaction with agreements would erode its already meager support in the steel regions. With a good grasp of the bargaining process, its negotiators were sufficiently astute to remind employers of the consequences of failing to meet demands. In such a way, a dynamic of negotiation came to characterize collective bargaining in the steel industry.

The CGT frequently presented wage demands that it knew owners would never accept. Such demands strengthened the negotiating position of the FO, which claimed that its positions were moderate. The employers also knew that they had to reach a "reasonable" compromise or it would never stick. Inasmuch as the CGT represented roughly 70 percent of French steel workers during the 1950s and together with the CFTC collected 90 percent of the votes in professional elections, the steel employers needed at least the implicit consent of one or both unions to keep workers on the job. Thus, to keep the CGT or the CFTC from calling strikes, employers needed to bargain "responsibly."

This interunion activity required considerable "learning" within the labor movement. The CGT needed to transcend its self-conception as the only working-class representative (*syndicat unique*). For the first decade after the war, both the FO and CFTC complained bitterly that the CGT pursued strike actions on its own or that it sought to bribe their militants.[44] They complained that the CGT did not accord them the legitimacy it gave to other organizations. This was more consequential for the CFTC, since the FO's blind anticommunism obscured the possibilities for joint action anyway.[45] The CFTC sought joint action with the CGT, which had an interest in cooperation, especially after its strategic turn away from sectarianism in the early 1950s. Mutual acceptance did not imply acquiescence to others' demands; rather, it meant limiting the damage of interunion rivalries and instead marshaling mobilizational energies against employers.

The bargaining process taught the CFTC both the limits of union bargaining power and the unreliability of the FO within the industry. The twenty years after the war radicalized the Lorraine CFTC. In 1945 its leaders believed that worker solidarity and conscientious bargaining would suffice to promote workers' interests. A new generation of leaders led by Eugène Descamps, Walter Paini, and René Carême infused the local and regional CFTC units with a sense of moral outrage at paternalist practices and the futility of relying entirely on the "shared consensus" of continuous bargaining. The union continued to attend bargaining sessions, but it also joined popular protests. Most important, it sought "unity in action" with the CGT despite its wariness of the PCF. It found the virulent anticommunism of the FO both distasteful and dysfunctional. The CFTC/CFDT wanted to present a united class front before the owners. That front might take the form of coordinated bargaining strategies or joint industrial conflict.

Both the CGT and CFTC perceived their activity during this period in largely negative terms. While the FO congratulated itself for its contractual gains, its two counterparts bemoaned the failure to realize programmatic objectives, at least into the mid-1950s. Their political agenda had been thwarted: nationalization was no longer a realistic goal and the steel barons set the terms of industrial change. Afterward, the failure to obtain their "hard" positions added to the disenchantment of the CGT and CFTC. Moreover, the feeling of being continually presented with *faits accomplis* dictated by the employers demoralized union militants.

The CGT and CFTC sought and failed to centralize collective bargaining during this period.[46] As a result of decentralized wage setting, there were important disparities in compensation. Workers with the same skills and seniority employed in two different shops within the same plant could receive hourly wages that differed by 20 percent. Disparities were even larger between plants and among steel regions. Moreover, regional negotiations spread organizational skills too thinly. Given the poverty of French unions, most of the full-time officials remained at the federation headquarters in Paris. Negotiations further strained union organization, requiring the participation of militants who needed time off for other responsibilities.[47]

The FO did not share this perspective. National bargaining threatened to reduce its role as first interlocutor because it might vindicate the negotiating capacities of its rivals. Given the relative weakness of the FO on the shop floor, it could not supplement its negotiating skills with the same level of militancy displayed by its rivals. The union was also infused with an American-style "business unionism" that gave plant-level job control priority over wider distributional concerns.

Given the organizational weaknesses of the unions, success in collective bargaining required a confluence of constructive negotiations and social mobilization. This meant abandoning internecine battles and establishing bargaining positions that could be achieved at the negotiating table or used as a mobilizing tool. Mobilization strategies developed as bargaining shifted from wage claims to job protection. The more focused the goal, the more likely that mobilization could give labor negotiators important resources in bargaining sessions. Here the CGT's aggressive pursuit of labor market demands meshed conveniently with new forms of social protest.

Changes of ownership within the industry, however, impelled an important evolution in bargaining. Because of centralized steel production in eastern France, the Lorraine accords gradually assumed a leadership role for other regional collective bargaining sessions. Union negotiators in other regions used the Lorraine terms as the basis for their negotiations. Since each federation sent negotiators from Paris to assist in collective bargaining, participants were cognizant of the details of agreements elsewhere.[48]

Moreover, development of the steel regions eroded the owners' paternalistic control over the housing, transportation, schooling, and religious practice of their workers. A new generation of managers perceived paternalism as archaic, favoring more regular work rules and more dialogue with the unions.[49] The mass production paradigm valued workplace stability more than "definitive" victories over working-class organizations. Consequently, employers bargained more willingly in the early 1960s than they had ten years earlier.

Conclusion

If the CGT and CFTC metalworkers' unions could congratulate themselves on anything in the first two decades after the war, it was maintaining their organizations. The two unions had to rebuild after repression and intimidation by employers in the late 1940s and early 1950s. Union publications carefully chronicled their organizational successes. The CGT, in particular, celebrated the courage of valiant militants in every dispute in which it was involved. This journalism, being mostly organizational self-promotion, also glorified the "culture of conflict" that was a substitute for the "culture of bargaining" lacking in French industrial relations.[50]

Until the mid-1960s, French unions in the steel industry sought to increase salary levels, regularize wage bargaining, and ensure the flow of investments into the industry. They shared with modernizers in Paris a concern for Malthusianism. The industry enjoyed prosperity, however, only after the mid-1950s, and profits resulted more from strong demand than competitive capacity.

Late modernization delayed the development of collective bargaining in the steel industry. Consequently, labor policy remained in the hands of the employers rather than being the object of negotiation. Employer control had retarded the changes in social relations that might have facilitated a reorganization of work patterns so as to speed the introduction of mass production. Thus, a vicious circle tied backward production strategies to confrontational social relations, and thus to employer attitudes that perpetuated those strategies.

The decentralized nature of French industrial relations would seem to invite plant-level collective bargaining, what Chris Howell calls "micro-corporatism."[51] Compared to federation staffs, local militants were better poised to engage employers, especially since they initiated the mobilization. Yet neither labor nor capital sought such a process in the steel industry. The employers avoided plant-level negotiations for fear of competitive bargaining with other employers, especially given the unions' capacity to mobilize locally. Instead, they sought to stabilize local labor markets through regional bargaining. Likewise, class-based unionism sought to reduce income differentials and to expand bargaining to the industry level. The two parties had an interest in divorcing the arena of bargaining from the locus of mobilization. Employer hostility combined with class-based unionism prevented fruitful bargaining at the local level, at least on its own. The industrial relations system in the French steel industry eventually joined with decentralized mobilization to force centralized bargaining. Given the financial precariousness of the industry, central bargaining virtually required state involvement, an obligation from which the government could no longer hide after the mid-1960s.

FIVE

The Creation of National Champions in Steel

Compromises over managerial autonomy haunted the industry. Steel investments had been poorly coordinated, and plant-level difficulties impeded the construction of a modern steel industry in France. Competitive difficulties in the 1960s were not simply cyclical but structural, and they resulted from the policy choices of the previous decade.

This chapter explores the attempts to reorganize the French steel companies in the mid-1960s. State industrial policy focused on whittling away at managerial autonomy in order to gain productivity. State intervention saved the industry from partial bankruptcy by injecting badly needed capital into the companies. The creation of national champions in steel promised to rationalize investments for a mass-producing steel industry.

The policy of national champions had its intellectual origins in the opening of the French economy. The Common Market, first in coal and steel, then in all industrial products, placed a premium on European competitiveness. Internationally competitive firms could be created either through rapid internal growth in a medium-sized company or the merger of smaller firms into a larger unit. Size would increase productivity through both economies of scale and development of an innovative corporate culture to spur competitiveness. With the *pantouflage* of civil servants into corporate leadership roles, national champions facilitated a greater state role in directing investment and organizing production.[1]

Political responses to market turbulence altered the pattern of interest intermediation in the steel industry. The trade association became a far more important actor in mediating relations between companies and the state. In particular, it substituted industry coordination for managerial autonomy by arguing that the coalescence of interests at the sectoral level could provide the environment for newly formed national champions. When the trade association assumed governance capabilities, however, organized labor entered the picture.

The unions gained in being consulted more regularly, although labor remained a junior partner. Unequal inclusion partially reflected debates within the labor movement over how best to confront the industry—whether to stress bread-and-butter wage demands versus comprehensive changes in property relations and industrial strategy. The temporary resolution of these debates enabled labor to coordinate strategy and mobilize so as to provoke the state to intervene. The unions used protest to compensate for their junior status.

Union strategy had evolved in the 1960s, and labor borrowed from precedent to demand state guarantees to cover the social costs of restructuring. The steel crisis of the early 1960s altered collective bargaining. Employment and wage bargaining assumed a national scope. This new pattern of bargaining removed wages from competition, as the unions had sought to do for the previous generation.

Unbalanced Growth in Steel

The simultaneous development of mass consumption and mass production came later in France than in Germany and the United States, as we saw in chapter 2. At the shop floor level, it was also delayed by a combination of investment choices and labor policy. Growth in the French steel industry was characterized by managerial inefficiency and a work organization more attuned to an earlier stage of history.

Steel is not a quintessentially mass production industry: it is not amenable to flow assembly, as with automobiles or consumer appliances, nor does it require extensive coordination among suppliers of interchangeable parts. It combines chemical reactions (within blast furnaces and steel converters) typical of batch production with mechanical transformations (rolling) whose techniques can be routinized—not to mention the procedures necessary to transport massive quantities of raw materials and the regular cleanings necessary for a pure product.

From a technical standpoint, steelmaking began to approximate the mass production model with a set of innovations in the postwar period. The continuous strip mill, which produced the steel used for automobiles and electrical appliances, revolutionized production by reducing the number of discrete steps in the rolling process. Oxygen steelmaking reduced the time necessary for the metallurgical process to eliminate carbon from iron. Computer processing in steel works and rolling mills eliminated traditional rule-of-thumb quality checks. These technological changes transformed steelmaking from a batch process to mass production.

Beginning in the United States and continuing in Europe, the steel industry came to be known for relatively high wages and low-skill work made possible by automation. Along with their counterparts in the auto industry, steelworkers could count on regular wage increases to propel them into the blue-collar middle class. This wage push was propelled by annual increases in productivity and facilitated by the economic rents gained through national oligopolies.

The patchwork of production facilities delayed the transformation to mass production in the French steel industry. An unwillingness to retire old plants brought average plant size below optimal levels and kept productivity low. In general, there was a mismatch between blast furnaces and steel converters, or converters and rolling mills or finishing facilities because of the large capital requirements of establishing an integrated mill and cyclical demand. Furthermore, French companies lagged behind their international competitors in efficiency, although the completion of Usinor-Dunkerque by 1963 boosted overall labor productivity. The drag of the less efficient plants was evident in production as well as productivity. Despite the addition of three "greenfield" plants before the mid-1960s—Usinor-Denain, Sollac, and Usinor-Dunkerque—the growth in tonnages was slower in France than in the ECSC's two dynamic steel producers, Germany and Italy.

The slow adoption of technological innovation enfeebled the industry. Using American designs from the 1920s, the mills at Denain and Sollac were smaller than optimal size for the 1950s. Still, because market demand for thin sheet took off—increasing sevenfold from 1950 to 1971—these mills eventually operated at near capacity. Nonetheless, Usinor had to build the Dunkerque plant because of Denain's inadequacies. The unwillingness to adopt oxygen steelmaking in the 1960s and continuous casting in the 1970s made it more difficult to confront international markets.

Investment choices influenced labor policy. The mixture of old and new technologies brought overstaffing, thereby accounting for poor labor productivity. It also produced serious bottlenecks. For instance, Denain's continuous strip mill started operations at 50 percent capacity for lack of steel to roll.[2] Sollac continued to use Thomas steel into the 1970s.

French steel managers consequently used long hours and speedups to increase production. More equipment might be added in older plants without a corresponding increase in personnel. Hours were long—forty-eight hours a week on average into the early 1960s—and mandatory overtime was frequent.[3] Work hours remained longer than those of ECSC competitors (see table 5.1). While the rate of decrease in working hours was slightly better than in Italy between the early 1950s and mid-1960s, it was much worse than Germany's record.

Working conditions could be dangerous as well as simply irregular. The unions complained incessantly about workplace accidents in the 1950s, although not all allegations are confirmed by reliable figures. One study conducted by the two employer associations in Lorraine showed that between 1951 and 1959 the accident rate fell by 29 percent and 46 percent in the departments of Meurthe-et-Moselle and Moselle, respectively. However, serious accidents (measured by fatality or permanent disability) increased by 39 percent and 23 percent in Lorraine over that period. The employers admitted that irregular hiring procedures and turnover in unskilled categories contributed to either inadequate training or carelessness in procedures.[4]

Workplace management was apt to be arbitrary in the French steel industry. Capriciousness was exacerbated by the technical hodgepodge of facilities, which demanded localized and even shop-by-shop differences in personnel procedures and work organization. Thus, wage set-

Table 5.1 Average Yearly Hours of Work, 1952–1966

	1954	1956	1958	1960	1962	1964	1966	% change since 1954
France	2,328	2,284	2,282	2,270	2,233	2,202	2,166	−6.97
Germany	2,324	2,286	2,094	1,973	1,917	1,951	1,927	−17.08
Italy	2,088	2,191	2,204	2,163	2,023	2,040	1,983	−5.06

Source: Author's calculations, based on Eurostat, *Iron and Steel: 1952–1982*.

ting was decentralized. The unevenness of technical change meant that changes in work organization were less widely diffused than in Germany or the United States. French steel relied on rule-of-thumb rather than the machine-dependent skills characteristic of the U.S. and German industries. Paternalism lingered partly because of the uneven rate of automation in the steel industry.

Steel mills in both the Nord and Lorraine operated or subsidized training centers to inculcate skills at all levels of the employment pyramid. While a young man immediately out of school could usually sign on for a semiskilled job entailing easy-to-learn, plant-specific skills, technical schools also taught apprentices who then graduated into skilled employment. By the early 1950s, Lorraine steelmakers were operating a regional "steel school" to train accountants, draftsmen, and supervisory personnel. (In keeping with the paternalistic ethos, young women could take home economics courses to become ideal homemakers for the young men in training.) The training also included a dose of company proselytizing.

The little documentation that exists on work organization in the 1950s and early 1960s suggests a widening rather than homogenizing of skill levels. Technical advances in blast furnaces (largely in scale) did not seem to alter work organization. Likewise, improvements made to either open-hearth or Thomas converters generally did not change skill needs. The new rolling mills of the 1950s, however, changed jobs for both blue-collar and white-collar workers. "Continuous mills" introduced office planning for individual job posts, whereas foremen in the old rolling mills "eyeballed" thicknesses, tensile strengths, etcetera.[5] The new mills used sensitive thermostats to replace the foreman's interpretation of metallic coloration. Workers complained that they could no longer "feel the metal."[6] The metallurgical precision introduced by automation dramatically reduced the level of unskilled workers and increased the demand for skilled workers, although semiskilled workers still dominated the shop floor.[7]

While the unions complained about employers' attempts to monitor, measure, and motivate work, the steelmakers were not particularly successful in imposing task management.[8] Anecdotal evidence suggests that streams of engineers performed time-motion studies.[9] The transfer of work planning to a central office, however, could not take place uniformly because of the wide array of technological systems within the steel complex. It also took time to eliminate or routinize ad hoc procedures. Finally, the combativeness of local Communist Party cells and poor relations between the steel companies and organized labor made for a tense social climate on the shop floor, which in turn delayed the introduction of Taylorist forms of work organization.

Unproductive facilities may even have bolstered skill levels in the work force. Unfortunately, reliable data is not available to compare skill levels cross-nationally. Given the success of its vocational and training system, Germany undoubtedly had a higher percentage of credentialed workers. A countertendency, however, was also at work: because of aging equipment and the need to rely on experience, French steelmakers may have had to depend more on their skilled workers. The goal behind the paternalist project, therefore, had a sound labor market logic. Apparently the division between blue-collar and white-collar work was roughly equal in France.

Salary levels were another indicator of antiquated work organization. The unions complained vehemently that wages were lower in France than elsewhere in the ECSC, while the employers remarked with equal bitterness that payroll taxes were the highest in the Community. Although cross-national data must be treated with caution because of problems of comparability, table 5.2 suggests that French employers exaggerated their burdens. They pointed correctly to the high payroll taxes they paid (the highest in the Community), but they did not calculate the low direct hourly wages.

Overall salary levels, however, were not the only bone of contention. The unions disputed the volatility of wages. Some evidence suggests that the French employers relied more on bonuses and piece rates than did their competitors in the Community, and this was accentuated by

Table 5.2 Total Hourly Labor Costs in the European Coal and Steel Community, 1962–1965
(in Belgian francs)

	1962	1963	1964	1965
France	59.95	65.05	69.70	73.86
Germany	75.11	79.31	84.29	90.44
Belgium	66.57	72.47	81.19	91.28
Luxembourg	74.71	80.96	86.23	97.34
Netherlands	73.36	78.95	87.28	97.81
Italy	60.85	71.69	78.68	80.44

Source: ECSC, High Authority, *General Report of the Activities of the Community*, various years.

the problems of older machinery.[10] The minimum wage may have played a role in pushing up wages for individual job posts. Because of the importance of gratuities and bonuses, however, wages varied considerably by job posting. Only with the continuous rolling mills and newer steel works and blast furnaces was there a move toward fixed salaries.

Employers insisted on setting wages locally. As a result, wages varied considerably by region. Consequently, steel was not a "pattern setter" in France as it was in other countries. An ECSC study found in 1965 that in France average hourly earnings in steel were exactly at the national manufacturing average, while they were substantially higher in the other five countries of the Community.[11] Steel employees could spearhead increases in mass consumption only with long workdays.

Thus, the first twenty years of the postwar period were characterized by erraticly organized production, harsh working conditions, long hours, and unequal pay. French steelmakers were not oblivious to the modern techniques of steelmaking and production design. They were behind, preferring to pay down debt rather than reorganize their production methods. The new managers who arrived on the scene in the 1960s changed organizational dynamics and attempted to rationalize production. The limited diffusion of mass production methods in steel, however, made their tasks difficult.

The Interventionist Temptation

The French state had conducted industrial policy in steel with arm's length policies under the assumption that leaving investment choices and managerial autonomy up to the firms would encourage a prosperous industry. When prosperity turned into deteriorating balance sheets, the terms of these calculations were called into question.

The European market was already suffering from surplus capacity in 1961: posted ECSC prices peaked in 1960–1961 and continued to decline over the following seven years. Low prices hurt French producers, since by the early 1960s they needed to renew plant and equipment. Aggregate productivity levels remained low by European standards. Only Usinor-Dunkerque, Usinor-Denain, and Sollac were truly competitive, and they all produced wide-band strip. Many of the smaller plants in Lorraine that made construction-grade steel operated at a loss, kept afloat with state monies. With their reluctance to incur debt, French steelmakers responded to the downturn by cutting back on their productive investments.

For the industry the challenge was financial, resulting directly from state price controls. During boom periods, companies could sell their high-cost steel, but profits were insufficient to reduce debt to manageable levels. During downturns, low prices made large segments of French steel uncompetitive. The strategy of using boom prices for debt reduction ignored the aging equipment, older technologies, and low productivity of numerous sites.

As the largest producer (20 percent of the market) and with the newest equipment, Usinor seemed best placed to ride out the storm of the early 1960s. The Dunkerque plant was producing 1.2 MT a year by 1964, and it continued to expand production. Yet the plant's construction had been expensive: by 1964 total debt exceeded 100 percent of sales. With less productive facilities, Lorraine-Escaut, Sidélor, and De Wendel could not hope to forge proactive strategies. Long-term debt for the entire industry was at 69 percent.[12]

Beginning in 1964, the trade association entered into negotiations with the state to relieve the beleaguered companies. The severity of the financial drain nixed the idea of continued voluntarism even for the MoI. The MoF assumed a skeptical posture: a condition for the comprehensive plan was a "reform of the industry."[13] This implied new mergers to consolidate assets. The Pompidou government explicitly excluded nationalization and *dirigisme*, although both policies had advocates in the MoF and MoI, but suggested that failing to relieve the debt burden might force a nationalization.

The CSSF reached an agreement in July 1966 to forge a comprehensive approach to industry problems by increasing liquidity, promoting mergers, lowering costs, increasing capacity, and retiring old plant. This Convention État-Sidérurgie would be accomplished through higher investments and more cooperation among firms, both of which were to be organized by the CSSF. The deal channeled 11,275 million francs (MF) to the industry—two-thirds of which was to finance productive investments, the rest to lower debt. The tab was financed with MF 2,700 of loans from the FDES, MF 2,700 of state-guaranteed bank loans, MF 1,000 of stock or bond issues, and industry resources for the remainder. Interest on the FDES credits was below market rates. Payments on the principal were deferred for five years.

The proposed mergers brought together assets into two national champions in basic steel (see figure 3.2, above). Usinor bought out Lorraine-Escaut, thereby gaining a foothold in Lorraine from which it could secure iron ore. De Wendel, Sidélor, and the SMS formed a cooperative to finance Sacilor, a new sectional plant in Gandrange (on a site abutting Sidélor's Rombas plant).[14] The venture concentrated long products in Gandrange and implied that redundant activities would be phased out, especially since the companies pledged to coordinate investments and production for their other plants. Coordination brought the companies together, and they merged into Wendel-Sidélor in 1968.[15]

Outside of crude steel, two other mergers took place under the auspices of the Convention. Two small producers of pig iron, the Société des Forges de Saulnes et Gorcy and the Société Nord et Lorraine, merged to create the Societé Saulnes-Uckange. In the specialty steel industry, the Compagnie des Ateliers et Forges de la Loire joined forces with the Société des Forges et Ateliers du Creusot to create the Societé Creusot-Loire. The new company was a major engineering firm with only half its sales attributed to steel: its subsidiary Framatome built most of France's nuclear power facilities.

The steel plan also detailed modernization measures in all areas of steelmaking; the technological changes met internationally recognized standards of optimal scale. The most important investments were directed toward oxygen steelmaking, which were given "absolute priority."[16] Gandrange received a basic oxygen steel works of 1.6 MT capacity with an increase to 2.6 MT envisioned. Usinor-Dunkerque would be enlarged to 3 MT capacity. Usinor agreed to modernize Denain by phasing out open hearth and Thomas converters, replacing them with oxygen converters. The Société Métallurgique de Normandie (SMN), an integrated steel mill outside of Caen owned by the Schneider group, received a small oxygen converter of 0.3 MT, to be doubled after 1970.[17] Overall capacity for the French industry was to increase from 23.3 to 25.8 MT between 1966 and 1970.

The state also pledged to help lower costs. The state-owned Charbonnages de France furnished coke and coking coal at competitive prices, and the railroad stabilized freight rates. The Ministry of Finance promised a revision of depreciation schedules to encourage investment. These provisions were designed to lower operating costs to European levels.

The Convention planned 15,000 employment reductions, two-thirds of them in Lorraine. The mergers and new technologies necessitated the closure of unproductive facilities. Sidélor's rail plant was closed, and the Saulnes-Uckange merger brought the closure of the Saulnes plant. The construction of Gandrange replaced more labor-intensive operations. Usinor's purchase of Lorraine-Escaut suggested that less productive sectional and plate mills in Longwy and Thionville might see less use than those in Denain and Dunkerque. In the Nord, Usinor-Louvroil closed its doors. Other plants saw partial shutdowns to make way for the new equipment.

The Convention was a coordinated effort for both the industry and the state. Previous steel policy had been formulated in the sectoral committees of the Plan, with ad hoc refinements for individual firms worked out with the Trésor. The Convention, however, was negotiated among a small group of people, including Jacques Ferry and representatives from the MoI, MoF, and Planning Commission. The arrangements implied a contractual obligation between firms and the Ministry of Industry. Thus, both production and investment became subject to state approval. Firm-level commitments were mediated by the CSSF. Ferry reinforced the already strong position of the CSSF.

The detailed interventions negotiated between each firm and the state limited managerial prerogative. Nowhere were the constraints more evident than in labor relations. Half the text of the Convention dealt with labor displacement. Given its experience in the coal and iron ore strikes of 1963 (to be discussed later), the government feared a social explosion and sought to provide some relief. The owners had to negotiate with the unions over the terms of industrial rationalization. The success of these negotiations was a precondition for the implementation of the Convention.

The crisis package increased the power and prestige of the CSSF and Ferry. Ferry positioned the CSSF as an intermediary between the companies and state agencies. With its connection to the GIS, the CSSF knew and influenced investment choices. The Convention inserted the trade association into firm-by-firm decision making in a manner unknown since Vichy.

The modernism of Ferry and the new managers fit in snugly with the perceptions of state actors. Ferry, in particular, led an industry known for its cautious, conservative style. He argued publicly that state-led adjustment was needed to modernize French industry, splitting with the employers' association (CNPF) over this issue. The "contractualism" of the Convention, he suggested, should be made general.[18]

As the first head of the CSSF who was not from a steel family, Ferry represented the new managerial class (typically trained in the *grandes écoles*) that later dominated French industry. Under his leadership, the CSSF played a "pivotal role in shaping the industry's strategy."[19] The collusion between the state and steel came at the expense of firms' autonomy at the same time that mergers gave individual firms greater marketplace power.

The negotiations surrounding Sollac's greenfield plant project at Fos-sur-Mer accented the duopoly in power-sharing between the CSSF and the state. Sollac had engaged with the state to construct a new coastal strip mill of world-class stature (optimal size was considered to be 6–8 MT).[20] The company favored a site near Le Havre for its proximity to major markets. The government favored the Mediterranean site of Fos for regional development purposes. Apparently, the actual decision to choose the Fos site over Le Havre was made without the knowledge of Sollac![21] Sollac's subsidiary, Solmer (Société Lorraine et Méditerranéene de Laminage Continu) was the centerpiece of a massive project to spur industrialization in southern France.

Increased state intervention in steel politics necessitated considerable damage control by the CSSF. Ferry recognized that the state would be motivated by both productivity and political concerns, and he accepted the idea that siting would result at least partly from the patronage it could offer. He worked with Sollac to negotiate FDES credits with a reluctant MoF, which hoped the project could be financed out of internal savings.[22] Ferry was aided by arguments from the MoI stressing that massive financing was needed both to build the Solmer plant and to rationalize production in Lorraine under the final year of the Convention.[23] Ferry later played an intermediary role between Wendel-Sidélor and Usinor to bring the latter in as coinvestor. He even worked with Finance Minister Giscard to entice the German firm Thyssen into a 5 percent participation. Usinor insisted that Ferry head Solmer.[24]

Nonetheless, in the late 1960s steel manufacturing was afflicted by the engineer's disdain for the bottom line—what McArthur and Scott call a "vocational" bias.[25] Production targets assumed sacred qualities. Aggregate tonnage served as the indicator for international competitiveness, and the yardstick was German steel. Both industry and state officials shared such sentiments, and management reshuffling failed to remove them. The fixation on targets was part of the development of mass production in steel.

The late 1960s, however, brought enormous enthusiasm for French steel. Markets were expanding. Prices were rising, increasing on average by 40 percent between 1968 and 1970. Capacity even became scarce in the late 1960s. Utilization rates shot up to 91 percent in 1969 and 1970.[26] Demand rose such that sales increased 65 percent between 1968 and 1970, during which time production grew by 16 percent. Pretax profits skyrocketed by 450 percent during these two years.[27]

Unfortunately, French steel could not take full advantage of this robust market. Productive investments scheduled under the Convention had started later than originally anticipated because individual contracts with firms were not negotiated until winter–spring 1967. Table 5.3 shows that investment projects did not start until late 1967 and early 1968.

The Lorraine group suffered longer from comparatively poor productivity. Sacilor delayed the Gandrange plant by two critical years in the midst of the boom and did not begin production until 1971. By contrast, Usinor stayed on schedule to enlarge steelmaking capacity from 1.7 to 3 MT at Dunkerque.[28] Wendel-Sidélor may have been the largest French producer by 1971—as owner of Sacilor, majority owner of Sollac, and with numerous older plants—but size masked serious weaknesses. The entire group (including its share in Sollac) had productivity levels 40 percent lower than Usinor's.[29]

The late start on investments had a perverse effect on restructuring: it delayed redundancies at some of the most unproductive plants (Micheville and Homécourt). The Convention had called for 5,800 job

Table 5.3 Productive Investments by French Steel Firms, 1963–1970 (in millions of francs)

<table>
<tr><th></th><th>De Wendel</th><th>Sidélor</th><th>Usinor</th><th>Lorraine-Escaut</th><th>Entire Industry</th></tr>
<tr><td>1963</td><td>174</td><td>139</td><td>378</td><td>163</td><td>1,610</td></tr>
<tr><td>1964</td><td>126</td><td>60</td><td>—</td><td>112</td><td>1,100</td></tr>
<tr><td>1965</td><td>117</td><td>48</td><td>—</td><td>93</td><td>875</td></tr>
<tr><td>1966</td><td>76</td><td>54</td><td>—</td><td>—</td><td>775</td></tr>
<tr><td>1967</td><td>100</td><td>76</td><td colspan="2">211</td><td>945</td></tr>
<tr><td>1968</td><td colspan="2">241</td><td colspan="2">333</td><td>1,345</td></tr>
<tr><td>1969</td><td colspan="2">182</td><td colspan="2">355</td><td>1,475</td></tr>
<tr><td>1970</td><td colspan="2">223</td><td colspan="2">588</td><td>2,060</td></tr>
</table>

Source: Company reports and the Chambre Syndicale de la Sidérurgie Française, various years.

losses between 1966 and 1970 for the three companies that became Wendel-Sidélor. Instead, there were fewer than 1,000 net redundancies.[30] For Lorraine as a whole, the industry had agreed to shed 10,000 jobs but actually cut only 62 percent of these. In contrast, the industry in the Nord met 91 percent of its obligations.[31]

Poor productivity drained Wendel-Sidélor of much-needed investment funds. Although the state formally lifted price controls in 1970, the effects were not felt immediately. World steel prices dipped in 1971 while raw material and labor costs skyrocketed (to be discussed later). Facing cost overruns at Fos and state credits from the Convention coming due, unable to raise funds on the stock market, and abandoned by one of its own principal stockholders, the group could not raise the self-financing necessary to sustain the project.[32]

The industry still perceived its problems to be financial and believed that state aid would enable companies to make the necessary investments to reach the economies of scale attained elsewhere. This argument was part of the "productivism" of the steelmakers. The quest for higher tonnages permitted the unbound optimism of the CSSF, which incessantly announced record-breaking levels for the industry or individual plants.

The CSSF negotiated another aid package in 1971 to help the ailing Lorraine industry so productivity increases could be passed on to Fos. The manufacture of sectionals was focused in Gandrange. Only those facilities that served steelmaking and finishing in Gandrange and Sollac remained open. Open-hearth and Thomas converters would be closed, as would small rolling mills. More than 12,000 jobs were lost. The plan resulted in a streamlined company that focused production of sectionals in Gandrange and flats in Sollac-Solmer.[33] The state would cofinance the redundancies through early retirements and interplant transfers—no layoffs would take place. Wendel-Sidélor legally reorganized under the name of its subsidiary Sacilor in 1973.

The working assumption, however, was that world markets would improve and French productivity goals could reach the European average. No attempt was made to approach the productivity levels of the Japanese or the Bresciani. The results were predictable given an uncertain market and high-cost production. The capital increases for Solmer that were supposed to have reached MF 1,700 before 1975 equaled only MF 400 in early 1972.[34] The delays increased construction costs by over 25 percent. The CSSF negotiated an additional MF 2,600 from the FDES in 1973, a sum that exceeded the annual budget of the state agency.[35] Wendel-Sidélor's position was sufficiently weak that it needed Ferry's talent to mobilize government support which, in turn, would allay the suspicions of the banking community.[36]

Twenty years after the inauguration of the ECSC, the French steel industry was dominated by two national champions—Usinor and Sacilor. The major investments went to the modern centers of production—Sacilor-Gandrange for sectionals, Usinor-Dunkerque, Sollac, and Solmer-Fos for wide-band strip. Worldwide demand perked up in late 1972 and continued for two years with consequent higher sales and prices. French production peaked at 27 MT in 1974.

The industry had an apparently secure relationship with Parisian ministries. The CSSF managed into the 1970s to secure stable funding levels. The more the state funded the industry, the more it became implicated in its success: steel ventures could not be allowed to fail. Ferry could play to this mutual dependence of state and steel. The critical position of the CSSF was based on the state's willingness to play the banker of last resort.

The period after 1966 increased government oversight of the industry but only through the mediating role of the CSSF. State involvement in investment did not entail control over the adjustment process: implementation remained with the individual firms. Oversight was left to the CSSF. Some projects received state funding but were never undertaken. Moreover, while the state bought redundancies, the firms continued hiring new workers, especially between 1968 and 1974. Intervention had not rationalized production.

The Short-Lived Glory of Mass Production

The French steel industry began to reverse the technical impediments to competitiveness with the Convention. It took years of state negotiations with individual companies and plants to provide the necessary investment to boost scale. Unfortunately for French producers, the pace of international technological innovation quickened after the 1960s. Not only were newly industrializing countries building new plants, but firms in the rich countries were investing heavily in machinery that would reduce input costs. In the face of market contraction, innovation was no longer a choice but a necessity to reduce costs and avoid being swamped by competitors.

The Convention set in motion the mergers necessary to achieve economies of scale. While concentration of assets preceded plant rationalization, plant size grew through the 1970s. Still, one study found in 1969 that only one-quarter of capacity was capable of producing 2 MT or more per year. In contrast, Germany had over half of its capacity in this range.[37] Yet, by the late 1970s, that difference had been obliterated. The construction of Gandrange and Fos as well as the enlargement of Dunkerque continued the trend of increasing plant size. Small plants were also retired. The investment in new machinery turned the tide on technological innovation. Tables 6.1 and 6.2 in chapter 6 show how French industry caught up to its rivals in the use of oxygen steelmaking and continuous casting—the two innovations that defined competitiveness in the 1970s.

The industrial strategy pursued both by the trade association and the state was to concentrate production in increasingly fewer sites—as opposed to the decentralization of production through minimills that characterized the U.S. and Italian industries. While investing in increasingly sophisticated technological systems lowered costs for raw materials and labor, it made organization of production much more rigid. Because it was necessary to operate at high capacity use, downturns in the business cycles and failing to fill orders could be disastrous. Therefore, work organization had to attain new levels of efficiency.

Economies of scale thereby changed the nature of work and remuneration. As can be seen in table 5.4, increased automation virtually eliminated unskilled workers from the shop floor and reduced the numbers of semiskilled workers. Mechanization did not require the large teams of simple human power that could be positioned to work in different parts of the plant under foremen's orders. Rather, the new equipment demanded machine-dependent skills. This shift stabilized immigration into the industry—by the early 1970s active recruitment was stopped.

The data on skilled and semiskilled workers does not fully represent the changes taking place within categories. For instance, there was a large flux of skilled workers. Those in older facilities were not easily moved to new machinery. A few were retrained, but others were eased out through early retirement. Otherwise, they were transferred to less qualified jobs but retained their job category.[38] In contrast, however, workers skilled in machine repair found their talents in greater demand.

Electricians, turners, fitters, and mechanics were trained in vocational schools and had not apprenticed under master craftsmen. Some of these workers would later be bumped into the category of technicians.

In the 1960s, employers started subcontracting the lower categories of the semiskilled. Manpower services became responsible for upkeep in the blast furnaces, managing the slag stocks, and so forth. These workers tended to take the more dangerous jobs, suffered the highest percentage of accidents, and did not benefit from collective bargaining accords.[39] (Partly as a result of this outsourcing, official accident statistics for the industry showed improvements.) Such subcontracting became a permanent feature of the new facilities. One study of Fos found six categories of subcontracting that ranged from janitorial service to maintenance work to engineering expertise. More than 20 percent of employees at the Fos site worked for external companies.[40] In 1989 over 17 percent of all jobs in Usinor-Sacilor were subcontracted.[41] The greater the number of external employees, the more difficulty unions had organizing the work force.

The numbers of other semiskilled job categories—crane operators, chargers, pushers—remained roughly the same. These workers were tied to individual machines. They frequently performed minor repairs in addition to their ordinary work, although formal training was minimal. While a significant percentage consisted of immigrants, they tended to be Belgians or Italians rather than North Africans.

The organization of work changed along with skills. Computer controls permitted greater monitoring of work. To the extent that the manufacturing ethos focused on production quantity, some form of Taylorism was envisioned by management. Managers sought to intensify work whenever possible in order to lower costs. For instance, at Fos workers were told explicitly that they were competing with subcon-

Table 5.4 Steel Employment by Job Category
(as percentage of total)

	Unskilled	Semi-Skilled	Skilled	Supervisors/ Office Staff/ Technicians
1966	10.6	40.9	28.8	19.7
1979	3.9	35.8	24.7	35.3

Source: CSSF, *La Sidérurgie Française*, various years.

tractors.[42] In the context of employment reductions elsewhere in the industry, such incentives were certain to speed up work.

Because of the late introduction of proto-mass-production techniques, however, managers could not implement task management. The continuing irregularity of work in an integrated plant—for instance, by the mid-1970s iron making was completely automated, but cleaning out the blast furnaces was still done by hand—made centralized time management impractical. Moreover, by the time of the second and third generations of computerization in the late 1970s, managers were more concerned with quality checks, inventory control, and production coordination than with time management.

These were also turbulent years for personnel management. Labor shedding after 1967 involved transferring redundant workers to more productive shops but brought instability to work teams. Early retirement eased workers out of unproductive plants but also distorted age pyramids and altered the mix of skills. While labor policy had to be more centralized to make these adjustments, the steel companies could not afford a single-minded focus on output. Rather, French steel needed increasingly to compete on the basis of metallurgical properties.

From the employers' perspective, working conditions improved dramatically in the years after the Convention. Worktime was reduced from forty-eight and a half hours a week in 1963, to forty-one in 1973, to less than forty by 1980.[43] Annual reports for Usinor and Sacilor, searching desperately to emphasize the positive in times of financial drain, glowed with lower accident rates, although they glossed over the complications created by subcontracting.

The steel companies also pointed to considerably higher wage costs. The negotiations after May 1968 dramatically increased wages, and wages increased faster in France than in major competitors (see table 5.5). Yet steel never gained the position as a wage leader that it held in other countries. The CGT published comparative wage data (based on a forty-hour week) in 1980 for the eight major metalworking industries and found that steel consistently ranked sixth in every major job category.[44]

The reversal in the technological position of the industry offered little gratification to employees. Employment instability instilled a collective psychosis. The unions pointed to promises made in the late 1960s about new investments and sustained prosperity not only in Lorraine but also at Fos that the companies were unable to fulfill in the 1970s when prices were low and international markets were soft. There were very few consecutive years of optimism for the industry after the 1960s.

Thus, the upturn in investments had a mixed effect in the French steel industry. The integration of production facilities into two national champions brought about the conditions for mass production. This increased productivity and provided higher wages. It also brought expanded employment, although only some of those jobs were highly skilled.

Labor's Perspective

Industrial restructuring via the Convention rankled French steelworkers. The CSSF refused any negotiations with labor over the terms of industrial change. Therefore, the steel bailout had been negotiated outside the auspices of the steel sector committee of the Plan. No negotiations took place over investment and production; the unions were simply informed of decisions. This not only confirmed suspicions of a "conspiracy to plan," but also played into the different strategic perspectives of the unions.

The metalworkers' federations from both the CGT and the CFTC had met frequently in the period after the CFTC-CFDT scission in 1964 to iron out joint demands over salaries, working time, retirement age, and union rights.[45] Such tactical cooperation also brought denunciations of the incipient restructuring of the industry. The unions' reaction was in large part seen through lenses developed over a previous generation of struggles.

For years the CGT had pushed a steelworker statute (*statut de sidérurgiste*). It was proud of the statutory provisions made by the Commu-

Table 5.5 Average Gross Hourly Labor Costs for Manual Workers in France, Germany, and Italy

	France (FF)	Germany (DM)	Italy (LIT)
1964	3.80	4.64	508
1970	6.37	7.46	884
1972	8.39	8.11	1,096
1974	12.58	10.15	1,586
1976	16.62	11.54	2,371
1978	20.34	12.76	3,169
1980	25.68	13.96	4,296

Source: Eurostat, *Coût de la Main d'Oeuvre*, various years.

nist minister of industrial production to protect coal miners in the nationalized pits in 1946. These provisions had been extended to iron miners in 1963. The CGT considered this protection mostly in terms of employment security, but the demand for a *statut* also served as an umbrella for other negotiating positions: monthly pay for all hourly workers, higher real salaries, cost-of-living adjustments, early retirement at sixty, a forty-hour work week, and expanded union rights.[46] These CGT demands were popular with the work force. Over the decade of the 1960s, the CGT moved toward positions once championed only by the FO, what one CFDT militant called "beefsteak" positions.[47] It pushed incessantly for a platform of "more"—more job security, higher wages, more purchasing power, and more benefits.[48]

If the CGT viewed industrial policy toward steel in terms of its impact on employment, the CFDT articulated employment growth with technological innovation. The union agreed with the CGT on employment and wages but suggested that the real problem facing French steel was productivity. Steel had developed in such a chaotic manner that production remained inefficient. Competition among firms had led to a duplication of investments. The industry required a carefully coordinated modernization, and only public control could rationalize installations. Nationalization needed to be complemented by coordination among firms, the government, the Planning Commission, and the ECSC.

Thus, the rhetoric of the CFDT combined nationalization and planning to increase production efficiencies. The CFDT saw nationalization in terms quite differently from the PCF and CGT. It argued that the Communist project simply replaced private capitalism with state capitalism, would result in autarchy, and ignored the institutions of the ECSC. On the contrary, the CFDT-FGM sought improved competitive capacity in a European and international environment over which it had little control.[49]

The CFDT-FGM connected nationalization of steel with democratized authority structures and concrete demands. Firms would be autonomous from the ministerial *tutelles*. Workers would be elected to sit on the board of directors, which in turn would elect the president. Union rights would be enhanced through union sections and stronger works councils. Nationalization would bring collective bargaining on a national scale in which unions and management could discuss all issues of pay and working conditions. Working time would be reduced to forty-two hours per week. Retirement would take place at age sixty. Salaries would be increased. Compensation would also be tied to profits. (In the optimism of the period, the CFDT did not discuss the possibility of losses.) For the CFDT, these issues linked the abstract notion of nationalization to the bread-and-butter demands of workers.

In the climate of the 1960s, however, this reasoning did not go over well. The CFDT-FGM made complex what many industrial workers saw as simple issues. Consequently, it did not always make the inroads it desired. A poll run by the Institut Français de l'Opinion Publique in Lorraine found that French steelworkers overwhelmingly felt the CGT did a better job in *protecting* immediate interests. In response to the question, "Which union defends the most effectively the interests of wage-earners," 46 percent of both blue- and white-collar workers suggested the CGT, while only 21 percent believed it was the CFDT.[50]

The Convention altered union calculations. The CGT-FTM saw state aid to the industry as expanding class exploitation, intensification of work, and monopolization of French industry. Rationalization, concentration, and modernization were seen as serving the sole purpose of increased profits and lower labor costs. This was the "result of the antisocial and antinational policies of the monopolies," policies "directed against the workers." The CGT-FTM opposed the public financing as a tax on the working population to "finance the monopolies." The task was to find ways to take the offensive against the monopolies and "Gaullist power." For this they needed to unite with other unions in opposition.[51]

The CFDT also opposed the Convention, but it argued that new investments were necessary for the competitive capacity of the industry. The new terms of international competition demanded modernized equipment, so the CFDT-FGM welcomed state intervention in the industrial sphere. However, it opposed the secretive nature of negotiations between the CSSF and the government. It also rejected the employers' position that the industry's debts were due to price controls; instead, they argued, poorly coordinated investments had created production inefficiencies. Intervention was the first step toward more successful investment control, control that would ensure the industry did not simply lay off redundant workers.[52]

The CFDT viewed the strident attitude of the CGT as undercutting effective opposition to restructuring. The CFDT-FGM opposed the process and feared layoffs, but it did not shy away from contact with participants. It initiated a series of discussions with Jacques Ferry and government officials in the spring and summer of 1966 to discuss the nature of the Convention. The FGM felt afterward that it had convinced Finance Minister Michel Debré to include employment protection.[53]

The difficulty of formulating union strategy in the 1960s was compounded by unequally distributed threats to employment. The inefficient plants of Lorraine and the Ardennes faced redundancies, while Dunkerque and Fos were gearing up. Even within Lorraine, while production at Villerupt was winding down, Wendel-Sidélor was constructing its Gandrange plant. While not exactly heady times, the late 1960s and early 1970s did witness considerable hiring by the employers and enthusiasm by state officials. Workers preferred to wait for the upturn in steel that arrived in the late 1960s.[54]

Different conceptions of industrial development and employment security within the CGT and the CFDT translated into different interpretations of the changes affecting the industry by the early 1970s. Both unions sharply criticized the project at Fos. The CFDT argued against the project on efficiency grounds: where were the markets for this new colossus? For the CGT, Fos meant the "destruction" of Lorraine. The transfer of production would mean less employment. Lorraine workers would "pay" for Fos with their jobs.

Mobilizing to Protect Jobs

The Convention gave the unions limited participation. They were authorized to focus on the "social" provisions of adjustment—clauses to protect the work force from losing their jobs. Under the terms of the Convention, the industry had to negotiate the appropriate language with the unions in order to qualify for funding. This led to a series of collective bargaining sessions between the unions and the employers in each steel region. The unions needed to translate limited inclusion into concrete gains for their constituents, a process that required a reevaluation of their strategic calculus.

By the mid-1960s, French labor was highly mobilized. The dilemma for the unions was making mobilization matter. The CGT and CFTC/CFDT had discovered that principled opposition to industry restructuring could not mobilize the work force. Both unions agreed that the twenty-four-hour strike actions called for December 14, 1966, for all Lorraine plants had largely failed. According to their analysis, the problem was communication: there was "a lack of awareness by the workers of our demands which, however, remain very global, too theoretical, and not sufficiently realistic."[55] In subsequent months, the two unions forged an agenda of demands that reflected joint concern over the level and nature of employment. These demands included issues of deskilling, transfers, early retirement at sixty, reduced work time, salaries, and bonuses.

A focus on bread-and-butter issues was accompanied by more cooperation in plant-level actions. Both unions wanted to avoid small, isolated strikes that employers could dismiss as irrelevant; they instead sought to coordinate actions over concrete demands that would garner attention. This "unity of action at the base" was accompanied by joint consultation between the CGT and CFDT at all levels of the two federations.[56] This combination of political opposition to restructuring—albeit from two very different directions—and tactical coordination brought the CGT and CFDT even closer together.

The steelworkers could point to the iron miners as an example of the fruits of militancy. A series of strikes rocked Lorraine's iron mines in 1963 over the terms of employment reductions.[57] The state organized a *table ronde* to bring the employers and the unions together to discuss competition, innovation, and labor adjustment. The tripartite meetings lasted two months and resulted in employment guarantees for all displaced miners. They set three important precedents that the steel unions sought to recreate. First, intervention came from the Ministry of Industry, not the Ministry of Labor, which emphasized the comprehensiveness of the negotiations. Second, the terms of industrial change were negotiated for an entire industry, not at the firm level.[58] Third, both employers and the state gave assurances that implied that workers had a de facto right to a job. Displaced workers would be offered either a possible transfer or payment for leaving the job.

The conflict in the iron mines is a fascinating story of labor militancy, collective bargaining, and state intervention. The iron miners had originally participated in solidarity demonstrations with the coal miners in March 1963. Job actions had been occurring in the coal fields for several months due to planned job reductions. The Pompidou government responded by requisitioning the miners under the excuse that supplies were low for home heating purposes. The depth of the coal miners' response—a thirty-five-day walkout and a willingness to confront President de Gaulle over public-sector wage guidelines—surprised the government. It also sparked the mobilization of iron miners who were facing similar problems. A week after the strike of coal (and iron) miners, the CGT miners' union organized its own march on Paris to protest the reductions in the iron mines. Assimilor and the Pompidou government promised the *table ronde* in exchange for an end to the industrial action. The *table ronde* met for two months between April and June 1963 and developed a set of "recommendations" for industrial and social policy. The miners were not completely satisfied with the terminology of the final agreement—"recommendations" were not guarantees. Moreover, there were indications that individual firms would not adhere completely to the language. When the mining subsidiary of the small steel company Société Saulnes et Gorcy, for instance, tried to lay off half its 550 miners in October, the CGT reacted by occupying the mine at Trieux for seventy-nine days. The occupation received widespread popular support in Lorraine. The CGT used the occupation to forward its own demands, which included work time reductions, reskilling, transfer, and the adherence to the *table ronde*. In January 1964 the company negotiated on precisely those terms.

The recommendations of the *table ronde* set important precedents for future negotiations in steel. The industrial policies had ranged from tariff increases on imported iron ore to increasing the demand for steel (and therefore iron ore) to regional diversification to create alternative employment opportunities. Assimilor agreed that any redundant miner had the right to a job in another mine (if available) or in the steel industry. Transfers would upset seniority structures, so transferred miners had the right to be trained for steelmaking skills. They received "lump-sum" payments to ease the transition. The cost of moving to a new job was covered by the employer. Tenancy in company-owned housing would not change. Thus, the social measures were comprehensive and designed to ease the workers' transition out of the industry.

The steelworkers' unions constantly pointed to the benefits received by the iron miners in their negotiations over redundancies in 1966–1967. After all, Assimilor had represented the iron companies, and the dispute had touched local nerves. (Many iron miners actually "transferred" out of mining and into steelmaking.) The Trieux iron mine strike in October 1963 had mobilized the local community. Most important, the comprehensiveness of negotiations appealed to the steel unions.

The Convention, however, differed from the *table ronde* in significant ways. According to the terms of the former, the industry had to negotiate a social plan with the unions to ease out workers without resorting to layoffs, but the state was absent from negotiations. In contrast, state officials played an important role in agenda setting and mediation during the *table ronde*. Moreover, the social plan that resulted from the Convention promised only to cover the terms of labor policy, while the *table ronde* discussed industrial and regional issues as well. The unions were relegated to their own "domain" of expertise in the steel industry.

Negotiations over the social terms of the Convention languished for several months before they were energized by a series of strikes. The Convention was announced in July 1966, but collective bargaining did not begin in earnest until autumn. The employers' association had called a collective bargaining session for early April. The press indicated that the FO and CGC were ready to sign an agreement governing the nature of the projected layoffs.

The industrial unrest began in an innocuous way when the CGT miners' union called for action after job reductions were announced for the iron mines. With miners blocking shipments, the steel owners threatened layoffs in the steel plants. Given the climate of uncertainty as ore stocks dwindled, steelworkers walked out, plant by plant, before they could be laid off. The entire Lorraine steel industry was paralyzed within a week.

The April 1967 strikes in Lorraine showed the mobilizational potential of the steel regions. Militants from the CGT and the CFDT coordinated information between plants and organized shop floor meetings to encourage industrial action. The CGT and CFDT cooperated in meetings with prefects and labor inspectors. They marched together in demonstrations. They coordinated positions for the collective bargaining sessions mandated by the Convention to deal with the proposed redundancies. Organizational consensus building within both federations reduced the divisions between union locals and their headquarters in Paris.

These strikes lasted for an average of three weeks, longer than any industrial action since 1948. Many observers later called these strikes the "May '68 of Lorraine." The 1967 strikes became town affairs. Steelworkers' wives demonstrated their solidarity. Local politicians took part in the demonstrations and denounced the employment reductions. Strikers received the moral support of the clergy. Communist local governments provided facilities for meetings, day care, and soup kitchens. While the Communists were active in the mobilization, they pursued actions favoring the positions of the CGT and not their own party agenda. Trade union unity facilitated mobilization and bargaining.

The 1967 strikes focused on labor market issues. They centered on the collective bargaining negotiations and the guarantees promised under the terms of the Convention. The CGT and CFDT agreed to call off the strikes only with the promise that their demands would be considered more carefully. Only at this stage did negotiations become serious. The agreement between the FNE and the CSSF to finance early retirement came in June. Final settlement was reached the next month.

The outcome of mobilization was critical for future politics in the steel industry. The social plan stipulated that no layoffs would take place. Employment was reduced through a combination of attrition, early retirement, transfers, and retraining. Moreover, Ferry promised to meet regularly with the unions to keep them informed of labor force changes.[59] The unions had succeeded in partially penetrating the armor that had protected the industry's cozy relationship with the state.

The Nationalization of Bargaining

In the period after the Convention, strategic analysis for both unions became increasingly grounded in employment protection. That required changing both the content and process of bargaining. Bargaining remained tied to labor market issues, and political statements receded into the background. Bargaining assumed a national scope in order to smooth out fluctuations for two companies that themselves drew from national labor markets.

The steel employers tried to keep collective bargaining at the regional level for the generation after 1945. Regional bargaining defused opposition to existing patterns of decision making. Union militancy, even in its "hard" forms, could be better contained within local labor markets. The CSSF fought ceaselessly to keep investment decisions and industrial planning a "private" concern for the companies and the state. The sessions in the steel sector committee of the Plan had developed into general forums for ratifying decisions made earlier between the industry and the state. Politicizing the issue risked upsetting the cozy business-state relationship of the 1950s, and it could degenerate into something worse—nationalization.

As the steel industry and its employment policies became more public and as the market deteriorated further, collective bargaining between the unions and employers' associations became more centralized. The social conventions for Lorraine and the Nord in 1967 followed previous bargaining patterns—that is, separate regional agreements were signed. Nonetheless, the government's insistence that state funding hinged on reaching agreement forced a similar responsibility onto all steel employers. As a result, the agreements for the two major regions showed remarkable similarities.[60] They were also signed within a month of each other.

The 1967 negotiations presaged increased nationalization of bargaining. The Convention foresaw solidarity among both competitive and noncompetitive plants on early retirement. Thus, the precedent was set for a competitive plant, such as Usinor-Dunkerque, to phase out older workers to make way for those transferred from other plants belonging to the same company, such as Usinor's plants in Denain or Longwy. Moreover, the unions met regularly with the national trade association to discuss how to implement the social plan to ease out workers. The unions participated on the tripartite boards to which redundant workers applied for transfers to the new plant at Fos in the early 1970s. The large number of transfers to Fos from Lorraine (3,000) testified to the nationwide scope of steel problems.

The employers bitterly resisted centralized bargaining. The regional employers' groups strove to retain control of the social agenda if only to prevent the unions from playing firms off against each other. For instance, conflicts between Assimilor and member firms took place over the use of "no-strike" contracts. Châtillon's plant at Neuves-Maisons tried to negotiate such a contract in October 1967 and even received favorable signals from the CFDT and FO. The proposed deal exchanged labor peace for a three-year employment guarantee and wage increases above those set by regional collective bargaining sessions.[61] The CSSF and Assimilor argued that such an agreement constituted a plant-level accord whose precedent could destabilize other Lorraine plants. As late as March 1977, Assimilor insisted that it still wanted to maintain the integrity of the local labor market. One official claimed in a collective bargaining session that since the state dealt with individual firms on financing, national negotiations were not needed. As if to soften his stand vis-à-vis the unions, he also argued that the difference between regions would exist in "form not substance."[62]

The employers' argument had its logic in previous patterns of collective bargaining. Several key issues had been resolved regionally and extended quickly to the national level in the years after 1967. The unions and the employers negotiated in April 1968 reducing the work week from forty-eight to forty-two hours. The agreement was partially foreseen by the Convention, which called for worktime reductions as one way to spread out jobs.[63] Still, the accord needed to be negotiated, and it was doubtful that Assimilor could find sufficient grounds for consensus with all five unions.[64] Usinor passed these concessions on to its employees in the Nord without the mediation of the Chambre Syndicale du Nord. The Grenelle Accords, which followed the strikes of May 1968, generalized these provisions nationally. The metalworkers formalized that generalization in December 1968 in an agreement signed with the UIMM.

Further, a collective bargaining agreement on sick leave (complementary to the social security system) was first signed in Lorraine and later copied by the rest of the metalworking sector. Here the results came from much earlier bargaining. The FO had negotiated separate agreements in 1956 with the Chambres Syndicales for the departments of the Moselle and Meurthe-et-Moselle. According to François Sellier, the UIMM negotiated a national agreement to avoid regional competition. The terms reached in the subsequent regional UIMM collective bargaining session in the Loire were less comprehensive in terms of the numbers covered but potentially more generous in compensation for insured workers. The UIMM sought to avoid union whipsawing and consolidated the sick leave system. The CFDT, FO, and later the CGT signed an agreement in 1967 that extended to the entire metalworking industry (1.7 million workers) the benefits obtained in the steel industry eleven years earlier.[65]

A national approach to collective bargaining could therefore be in the interests of both employers and unions. This was evident in the agreement on monthly wages. The negotiations on *mensualisation* led to changing the system of remuneration from hourly to monthly for most job categories.[66] Neither the metalworking trades nor the steel industry was in the forefront of this reform. Still, the negotiations set a precedent for both steelworkers and metalworkers, since they took place in Paris and transcended the regional bargaining demanded by both the UIMM and the CSSF.

Employment bargaining in the steel industry became national in scope not only because of precedents set in previous bargaining sessions, however. Union pressure to centralize collective bargaining became irresistible with the necessity of plant-level restructuring. Jobs had to be shed, and the state demanded that the social provisions be negotiated between management and labor. Since those provisions precluded layoffs, the only mechanism to reduce labor would be to shift employees between plants. This required the development of a national steel employment policy.

Conclusion

New blood helped change industrial politics. Not only were new managers willing to relinquish a certain amount of decision making to the trade association, but they combined assets to achieve the economies of scale necessary to remain competitive in increasingly open markets. The CSSF, in turn, proved a capable defender of industry interests, pressing the companies to modernize and negotiating with organized labor.

The deterioration in the industry's health politicized decision making regarding the adjustment process. State industrial policy increasingly became the focal point of union protest, for the government intervened at the hint of social turmoil. Intervention undermined the social bases of paternalism in the 1960s and eroded the employers' autonomy over personnel decisions. The Convention and subsequent state policy insisted that workers be protected from the nature of industrial change.

State intervention altered labor politics in the industry. The unions finally gained a limited measure of inclusion. More important, they gained a lever in labor policy. The state had an interest in maintaining social calm, and the unions had the capacity to mobilize. This combination brought employment guarantees to steelworkers.

SIX

The Politics of Industrial Decline

The steel market changed dramatically in the generation between the formation of the ECSC and the creation of national steel champions. As a capital-intensive sector, steelmaking is vulnerable to fluctuations in the business cycle; when sales dip, capacity use drops, an enormous amount of capital becomes unproductive, and companies lose money. In the past, French industrialists had survived temporary downturns by postponing modernization, increasing debt, and asking the state for financial assistance. Faced with an ever less competitive steel sector, the state intervened to create a mass-production industry.

The declining demand for steel involved lasting changes, not merely short-term fluctuations. Shifts in consumption patterns reflected a long-term substitution of plastics for steel in products ranging from radios to autos. Newly industrializing countries rapidly added capacity, satisfying domestic demand, and sold at below-average prices in rich markets. Like those in France, producers in other rich markets had planned new capacity in the late 1960s; with a minimum of four years required between planning and production, these new plants were coming on line in the mid-1970s. By 1974, steelmakers in the OECD world faced overcapacity, uncertain demand, and long-term decline.

The depression in international steel markets after 1974 hit French steelmakers particularly hard because product market adjustment strategies had not anticipated these structural changes. Under the national champion policy, sheer tonnage had assumed patriotic meaning, and more profitable, less showy ventures were shunned. The French industry had maintained maximum productive capacity, regardless of demand, and failed to retire unproductive plants and machinery. If steel markets had expanded again as experts predicted, French industrialists might have survived to produce steel, albeit at lower than average profits and high unit labor costs. The longer the downturn lasted, however, the less room the steelmakers had for maneuver. The state was too implicated in the steel industry to let it slide into bankruptcy, yet it could not prevent disaster without even deeper involvement. The severity of the crisis weakened the trade association and justified full-scale product market intervention by nationalizing the steel companies.

State intervention to avert bankruptcy created a labor market crisis in the regions that depended on steel for employment. The tensions that characterized the French labor movement intensified in response to the radical restructuring of steel production. Through a series of dramatic strikes and decentralized protest actions, workers expressed deep moral outrage over the job losses that shook their communities and contradicted the optimistic promises of the steel industry and the state. Union organizers and leftist party leaders struggled to channel the mobilization into concrete gains, but attempts to link local labor issues to broader political positions reduced the number of workers who would take to the streets. Popular mobilization forced the state to include labor representatives in the social component of industrial adjustment. Access to the negotiating table, however, created new dilemmas for labor leaders. While they recognized the need for union solidarity, they had difficulty deciding between *oppositional* demands for increasing worker benefits or *propositions* for strengthening the industry's employment and competitive capacity. As the CFDT found more effective ways to combine militant mobilization and successful negotiation, it was able to alter interunion relations. Yet the realization of labor's demands for inclusion in national-level negotiations and nationalization of the steel industry came more as side effects of product market adjustment than as rewards for effective union action. These victories proved bittersweet in the midst of widespread industrial decline.

Market Failure and the Statist Response

The decade after 1974 redefined the adjustment pattern in the French steel industry. After massive infusions of capital, the state took incremental measures to ensure that the companies made the adaptations necessary to increase productivity and restore profitability. Because of the weakness of the steel market and the legacy of investment choices, these measures eventually resulted in the government's takeover of assets and the rationalization of production, thereby terminating the industry association's mediating role and inaugurating a new phase of industry governance.

Most European steelmakers were slow to spot the nature of the forces affecting their industry: after all, consumption changed irregularly over the decade. In France, the Seventh Plan offered two working hypotheses of predicted growth in demand for steel between 1974 and 1980—1.9 percent and 3 percent—yet instead, market demand *declined* by an average of almost 3 percent in that period.[1] Industry capacity increased from 30.5 MT to 32.7 MT between 1974 and 1978, while production decreased from 27 to 22.8 MT. The decrease in capacity use forced companies further into debt.

Given the market downturn, the industry had two possibilities—either to close down immediately the unproductive facilities that put a drain on company finances or invest in new facilities to increase productivity. The latter strategy was constrained by the companies' debts. In 1974 the French industry was four times more indebted per sales unit than its German counterpart.[2]

The French industry responded to low capacity use and high capital costs by belt tightening. Sacilor delayed investments in Lorraine—continuous casters, oxygen converters, and an internationally competitive blast furnace at Joeuf. The second phase of Solmer was shelved indefinitely. Usinor also found itself in a weak position. In December 1974 it tried and failed to acquire Creusot-Loire, which would have immediately created a modern integrated metalworking company. The deal would also have hived off its unproductive Lorraine activities (from the earlier merger with Lorraine-Escaut). Consequently, the company was saddled with additional debt and Lorraine facilities that desperately needed restructuring.[3]

These foregone investments would have increased productivity and lowered unit labor costs. As can be seen in tables 6.1 and 6.2, the French industry remained behind its rivals in adopting new technologies until the late 1970s. Sacilor's restructuring plans were not yet finished, and while productivity improved, it was still inferior to Usinor's, not to mention the German average.[4] Sacilor produced only 65 percent of its steel in oxygen converters—less than Usinor's 75 percent. French productivity remained the lowest in the six-nation European Community.[5]

Yet a strategy of closing down unproductive facilities was foreclosed by government labor policy. As part of his government's countercyclical economic policy, Prime Minister Jacques Chirac forbade any employment reductions in 1974–1975. In exchange, the MoI consented to MF 1,500 in loans from the FDES to meet interest payments and to engage in countercyclical investment. The state and industry were paying the price of their clumsy labor market tools (see chapter 2).

If the market downturn had been temporary, such measures might have helped the industry bridge difficult times. Yet the continued weak-

Table 6.1 Adoption of Oxygen Steelmaking
(percent of crude steel production)

	France	West Germany	Italy	United States	Japan
1960	0.5	2.5	—	3.4	11.9
1965	13.1	19.1	22.0	17.4	55.0
1970	29.0	55.8	31.5	48.2	79.2
1974	58.5	68.8	45.7	56.0	80.9
1979	79.7	76.1	42.0	61.0	76.4
1982	80.9	80.9	47.4	60.8	73.4

Source: OECD, *The Iron and Steel Industry*, various years.

Table 6.2 Adoption of Continuous Casting in Selected Countries, 1965–1985
(percentage of total crude steel output)

	France	West Germany	Italy	ECSC TOTAL	United States	Japan
1965	0.6	2.0	0.7	1.1	—	—
1970	0.8	8.2	4.2	4.2	—	—
1973	7.3	16.3	16.1	9.4	6.8	20.7
1974	10.2	19.4	21.7	12.6	8.1	25.1
1977	23.7	34.0	38.5	25.4	12.5	40.8
1979	29.7	39.0	46.4	30.9	16.9	52.0
1980	41.3	46.0	49.9	39.2	20.3	59.5
1982	58.5	61.9	58.5	52.8	29.0	78.7
1984	67.1	76.9	73.2	65.6	39.6	89.1
1985	80.7	79.5	78.6	71.4	44.4	91.1

Source: OECD, *The Iron and Steel Industry*, various years.

ness of steel markets made debt servicing unbearable. By the end of 1975, long-term debt equaled 98 percent of sales at Usinor and 77 percent at Sacilor. In 1975, 1976, and 1977, Sacilor lost a total of MF 4,745, while Usinor lost MF 4,526. By the end of 1977, long-term debt as a percentage of sales stood at 159 percent for Usinor and 150 percent for Sacilor. *Fortune* referred to Sacilor and Usinor as the most indebted large companies in the world.[6]

The state responded initially to the weakening market in predictable ways. The MoI encouraged the concentration of steel production outside Usinor and Sacilor-Sollac. Three small firms—the Société des Hauts Fourneaux de la Chiers, the Compagnie des Forges de Châtillon-Commentry-Biache, and the Société des Aciéries et Tréfileries de Neuves-Maisons-Châtillon—grouped together under the name of CICC (Compagnie Industrielle Chiers-Châtillon). The new company had 2 MT of capacity (0.3 MT in specialty steels), a niche in high-quality cold rolled sheet, an array of profitable product lines, and less debt in 1977 than the big two (long-term debt equaling only 82 percent of sales).

Yet state actors focused attention on the fate of Usinor and Sacilor. The CSSF had tried unsuccessfully in 1975 to raise prices by implementing Article 58 of the ECSC—a declaration of "manifest crisis" that would enable the Commission to organize the market, reduce third-country imports, and raise minimum prices. Ferry also approached the government in December 1976 for additional credits to rescue the companies.

The April 1977 Plan Acier challenged the integrity of the industry, revealing "the first rupture in the steel consensus."[7] An interministerial committee (from the Trésor, MoI, and Ministry of Labor) saw the industry's request for additional credits as "irresponsible" and sought more public control.[8] As a condition for a new infusion of aid, Usinor and Sacilor had to deposit MF 550 of their own funds and MF 100 of stock with the Caisse des Dépôts. The government also insisted on the three-member commission to watch permanently over the firms' commitments. Moreover, the holding companies (DNEL for Usinor and CLIF for Sacilor), which had been profitable in spite of their subsidiaries' losses, were required to merge production and nonproduction assets to secure additional funding.

This rescue package injected MF 1,300 of FDES credits into the industry. Another MF 500 came in January 1978 under the same auspices. The companies promised to reduce 17,000 jobs by further rationalizing production, which the state would support at an estimated cost of MF 1,000.[9] The package included further concentration of assets. The SMN received oxygen steel converters. As part of the move toward downstream diversification, it merged in September 1977 with the Chantiers de France-Dunkerque, a shipbuilding company also owned by the Schneider group to become the Société Métallurgique et Navale Dunkerque-Normandie. Cockerill-Rehon (outside of Longwy) contracted to build an oxygen converter in 1978, and the Société Nouvelle des Aciéries de Pompey (SNAP) received a roughing mill.

These measures were too little, too late, to stop the market deluge. The state had again accepted the industry's optimistic forecasts, and capacity was not reduced. Instead, production declined in 1977, and orders for 1978 showed further erosion, prices remained low, and firms' finances continued to deteriorate. The 1977 package tried to reduce indebtedness to 70 percent of sales, but the FDES credit only helped pay off credits coming due.

The increasingly tentative relationship between the industry and government disintegrated by spring 1978, when the latter took over the companies. Yet the government needed the political space for maneuver. As the crash assumed even larger proportions and as the industry became an even larger embarrassment, creditors were told to wait for the next electoral outcome, the March 1978 legislative elections.[10] Prime Minister Raymond Barre immediately announced another "rescue" package, the fourth in twelve years.

The Plan de Sauvetage sought to defuse once and for all the financial crisis facing the industry. It rejected bankruptcy, as recommended by liberals, and nationalization, as suggested by the Left. The former would have been unpalatable to the Center-Right's constituencies. Nationalization, while contemplated within the state bureaucracy, would have been politically unwise for the Barre government. Instead, the state took control of the industry *without* ownership and arranged for repayment of debt.

The state took five steps to improve the financial position of the steel industry. First, the shares of both companies were devalued—Usinor's by 33 percent and Sacilor's by 50 percent. Second, creditors transformed MF 2,000 of their loans to the two companies (10 percent of their total debt) into shares in two holding companies. These holding companies assumed ownership positions, increasing the base capital of both firms. Third, more "good faith" measures required transferring two profitable ventures to the new holding companies. The CLIF transferred the Saar steel mill Dillingen Hütte to Sacilor, and the DNEL relinquished Vallourec, the second largest European tube producer, to Usinor. Fourth, credits from the FDES (MF 8,000) and the GIS (MF 13,000) were converted into "participatory loans," maintaining a fiction that the industry could buy out its creditors when it "recovered."[11] Fifth, the banks, holding MF 9,400 of debt, converted MF 600 into stock (30 percent of new holding companies) and limited their interest payments to a fixed annual sum of MF 80 for five years. In addition, a new state agency assured the small bond holder that individual debt would be respected. The Caisse d'Amortissement pour l'Acier (CAPA) promised to watch over company payback schedules and take direct responsibility for repayment. With its own funding, the CAPA could pay off the MF 13,000 in bonds coming due and wait out the deferred period steel companies were given to pay off their obligations. These financial measures promised to reduce indebtedness (and service payments) to German levels. Equally important, they enabled the state to purchase the control it had lacked in formulating previous policy.

The Plan de Sauvetage also contained two important industrial measures. First, assets were further concentrated. Usinor purchased the Société Châtillon-Neuves-Maisons from CICC and Cockerill's plant in Rehon, outside Longwy.[12] Sacilor-Sollac bought the specialty steel company SNAP. Second, the Plan reduced capacity by 5 MT.[13] Usinor closed the iron and steel works at Denain, leaving open only the large band strip mill. It also closed the steel works in its Chiers plant in Longwy as well as Usinor-Longwy (the old Lorraine-Escaut plant), thereby concentrating its Lorraine steel production in Rehon and Neuves-Maisons.[14] (It also promised to build a minimill in Thionville.) Industrial restructuring was less dramatic for Sacilor-Sollac. Sollac received new oxygen furnaces and two continuous casters. The Thomas steel capacity at Hagandange was shut down.

Rationalization of production brought 20,000 redundancies. For the fourth time in twelve years, the state purchased the employment reductions with social guarantees, especially important given the worker violence that accompanied the announcements (to be discussed later). The costs of the social measures in 1979 amounted to MF 11,000—the most expensive and comprehensive that any French industrial sector had experienced.[15]

To gain the political space to carry through these measures, the state needed to divorce itself completely from the CSSF. Prime Minister Barre claimed to have been duped by the trade association.[16] The government therefore developed its policy entirely outside industry channels. Ferry was consulted only in his capacity as head of the GIS—one of the creditors—and he was excluded from substantive negotiations. The presidents of Sacilor and Usinor were never consulted. The Trésor assembled the industry's major creditors and negotiated the financial measures. The technical services of the MoI formulated the accompanying industrial measures. The state then removed the heads of Usinor and Sacilor and installed leaders of its own choosing. It also forced Ferry's resignation as president of the CSSF. The departure of Ferry formally ended the CSSF's mediating role.

This political settlement was facilitated by changes that had taken place within the industry. The 1976 and 1978 mergers had already changed interest representation, as Usinor and Sacilor had less need of the "good offices" of the CSSF. The depth of the company's losses gave the industry little room for bargaining.

Relegating the CSSF to outsider status (mostly statistical work) placed the responsibility for the steel industry squarely on the shoulders of the French state. Unambiguous control over adjustment policy did not necessarily entail success. The new policy was to target market niches for individual steel products, thereby breaking with the industry's productivism. The new emphasis would not be on more but where. Yet financial decisions outweighed production ones, for Barre's goal had been to relieve the drain on the company finances. Production was cut back for financial reasons, not in relation to market strategies for individual products. As a result, production bottlenecks increased. For instance, the closure of steelmaking in Longwy increased Usinor's costs. Neuves-Maisons's steel converters were to be modernized to supply Longwy. Transportation costs for the 100 km. to Longwy could have paid for the oxygen converters that had been planned for Usinor-Longwy.[17] Similarly, the shutdown of steel capacity in Denain forced Dunkerque to run at maximum capacity use during normal periods of demand.

Moreover, the French state again stopped short of constraining behavior, as it followed many of the old practices of the CSSF. The MoI did not insist on production integration to accompany the rationalization measures. It did not merge specialty steels with crude steel production, although this would have both encouraged the companies to produce higher value-added products in the long run and sounder balance sheets in the short run. Likewise, the companies did not dramatically revise recruitment policies in spite of redundancies. The passivity of labor policy (only reducing the numbers in the work force) prohibited a careful evaluation of which workers were performing which tasks. The Barre measures gave the quasi-nationalized companies considerable autonomy with their new financial solvency, but it would return to haunt them.

The Barre plan failed to improve financial performance. As late as spring 1981, the government needed to extend further subsidies to keep the companies afloat.[18] French steel was restructured at the time of the second oil crisis. International demand plummeted, and newly industrialized countries made inroads in third-country markets. A price war erupted within the European Community. Without downstream activities, French steel was more exposed to market fluctuations than its competitors. Productivity had improved dramatically, but it still lagged behind other European competitors. Only in the early 1980s would continuous casting in France achieve the levels of Germany (see table 6.2).

The Socialists fell into the same trap of productivism that had ensnared both the industry and previous governments. François Mitterrand had promised in the 1981 presidential campaign that no jobs would be lost in the steel regions. The Left argued that more investments in capacity would enable the industry to master the market. The goal of Socialist industrial policy was to "reconquer the domestic market." The initial plans for steel reflected such thinking. Reconquering domestic markets meant producing a wider range of products that would give a "new coherence" to the industry.[19] The Socialists would reverse policy and ensure that "there are no condemned sectors."[20] The new government perpetuated a norm that had characterized adjustment in steel since 1950: infusing new money into the industry promised prosperity.

Overt nationalization was the first Socialist policy move toward the industry. De facto ownership became de jure: the "participatory" loans were converted to stock. By nationalizing the financial sector and other actions, the state owned 86 percent of Usinor and 95 percent of Sacilor. The Left also removed the two CEOs and replaced them with Raymond Lévy at Usinor and Claude Dollé at Sacilor.

The Socialists encouraged further concentration of production, combining specialty with crude steel production. In 1981 Usinor negotiated a complicated arrangement with Creusot-Loire by which the latter created a subsidiary to manage three of its specialty steel plants, and Usinor bought a majority stake in the new company, the Compagnie Française des Aciers Spéciaux.[21] CFAS then purchased the Société des Aciers Spéciaux de la Chiers.[22] In 1983 Usinor purchased most of the other plants of Creusot-Loire.[23] The consolidations enabled Usinor to become a major European competitor in stainless steels and high-quality steels. Sacilor also took over Ugine Aciers and the Société des Aciers Fins de l'Est, giving it a complete line of specialty steel products.[24] State subsidies promised to cover at least half the cost of reorganizing production at Ugine. Sacilor also bought the SMN in Normandy and a minimill in Montereau, south of Paris. By combining crude steel production with higher value-added activity, the state started the process of downstream integration and greater insulation from market forces.

Still, the short- and medium-term prospects of the industry needed attention. The long recession of the advanced industrial countries continued to depress steel prices. The new government commissioned four reports to assess the status of the industry, only one of which has been released. The Judet report analyzed the future markets for French steel products, using three hypotheses on the future of domestic and international markets.[25] The Socialists chose the most optimistic forecast to frame steel policy.

The 1982 Plan Acier—the fifth in sixteen years—directed financial grants to more investments and new machinery. It was also intended to cover losses and reduce debt. It aimed to produce 24 MT by 1986 but simultaneously reduced capacities by 5 MT. No plant closures were involved, although another 6,000 redundancies at Sacilor and 4,000 at Usinor would take place. The new government committed MF 20,700 from 1982 to 1986 to modernize the industry with continuous casters, modern rolling facilities, and renovation of various facilities. It focused production on the four cores—Gandrange, Sollac, Solmer, and Dunkerque.[26] Both Usinor and Sacilor created subsidiaries to intervene in the steel regions to promote new investments and new jobs—recognizing the dependence of Lorraine and the Nord on steel. Iron ore mining and steel production were finally decoupled in the 1980s.

Most observers characterized the Plan Acier as unduly optimistic. The goal of 24 MT was more an act of faith than a correct reading of the market, since production had not reached that level since 1974.

French steel still had a patchwork of production facilities, although older equipment was being scrapped. The French companies lost a total of MF 26,539 between 1981 and 1983. Production stagnated at 18 MT, and each ton produced below the Plan's forecasts (22 MT for 1984 and 24 MT for 1986) lost 1,000 francs.[27]

The combination of surplus capacity and state intervention continued to politicize adjustment. The typical promise of future profitability led to future embarrassment when the market did not perform as expected. It provided a focus for anxiety over the future of steel regions. It is not surprising that steel absorbed the lion's share of state subsidies to all public firms between 1981 and 1985. Nor is it surprising that the government continued the generous payoffs to labor to purchase the necessary redundancies.

State control made regional trade-offs difficult to avoid. The decision to close the strip mill in Denain and not Sollac seems to have been made in an effort to maintain employment in Lorraine, especially since Denain had similar facilities and was closer to port facilities.[28] Likewise, the purchase of the SMN added surplus capacity to Sacilor's product lines and was likely to provoke debate between Lorraine and Normandy.

In the decade from 1974 to 1984, ownership and control were merged for French steel. This altered the pattern of industrial governance by displacing the trade association. Nationalization altered the locus of power in industrial politics, but it did not solve the problems of product market adjustment.

Mobilization to Confront the Employment Hemorrhage

Massive state intervention to restructure the steel industry had two major effects on labor. First, it created an employment crisis that stunned both the unions and their constituencies. French steel workers had expected the restructuring implemented under the 1966 Convention to restore the industry to health, so the dramatic job losses of the mid-1970s stirred both anger and fear. The guarantees in the 1967 employment negotiations had increased union membership, so labor could wield organizations more readily to protest job reductions. And since restructuring threatened a way of life, protests enjoyed widespread support in communities throughout the steel region. Second, state control increased the value of political action to influence industrial policy. Electoral success for leftist political parties would ensure that policy makers shared labor's strategic vision. Industrial centralization also meant that effective negotiation could bring benefits for steelworkers across the country. Unions thus faced the crucial dilemma of how to translate social mobilization into the political power needed to influence the terms of adjustment.

French steelworkers in 1973 had good reason to be optimistic. The national champions were breaking production records—tonnages peaked in 1974. Industry investments in the 1960s brought more sophisticated machinery that demanded new job skills. Technological change—oxygen steelmaking and computerized controls as well as continuous casting—eliminated most unskilled jobs and redefined skilled workers into technicians. French steel seemed once again on the cutting edge.

The nature of workplace representation was also changing. Union membership was increasing, with the decline of employer paternalism, the reforms of May 1968 that prevented employer opposition to shop floor union sections, the simplified managerial structure of national champions that facilitated organizing, and the unions' success in mobilizing the 1967 strikes. The balance of union forces, however, shifted during the 1970s. While the CGT had always been stronger among skilled and manual workers, the CFTC/CFDT constituencies were comprised of white-collar production and clerical workers. Yet industry modernization favored technicians and white-collar workers more than production workers. Thus, modernization favored the CFDT. This relative decline of the CGT (the FO remained largely on the sidelines) did not reduce shop floor mobilization. On the contrary, the dynamic of competitive unionism may actually have increased shop floor demands, as no individual union organization dominated and each remained competitive by forwarding demands on behalf of existing or potential members.

Labor's initial response to the downturn in demand for French steel was to deny that a genuine crisis existed. The CGT and CFDT met frequently to forge common positions.[29] Even while national and regional newspapers were reporting record losses for the companies, the unions published numerous tracts that denied the problem of surplus capacity. For the CGT-FTM, the so-called crisis was organized by the owners in conjunction with the state in order to fabricate higher profits. According to the CFDT, the owners sought "to mask the record profits never before realized in the steel industry in order to not respond positively to the demands of the workers."[30] In the face of lower orders for the companies, lower production for the industry, and lower prices in world markets, the CFDT suggested that the steelmakers were organizing "an artificial steel crisis."[31]

The refusal to recognize market trends can be linked to the national electoral ambitions of the CGT and the CFDT. The CGT in particular became infatuated with the possibility that the Left could assume national political power. The alliance between Communists and Socialists, cemented by the Common Program of Government in 1972, had been gathering steam throughout the 1970s. For the unions, implementation of the Common Program entailed the long-sought goal of nationalizing the steel industry. If the CFDT had its way, nationalization would be accompanied by a "democratization" of the steel companies. Such hopes had been nourished at least since Mitterrand's narrow defeat in the 1974 presidential election, leftist victories in parliamentary bi-elections, and the Left steamroller in the 1977 municipal elections.[32]

These hopes spawned a strategic orientation at the confederal level that emphasized political work over labor market actions. Frustration with the Right's long tenure in power led the CGT and CFDT to put their strategic eggs in the basket of public control. When redundancies were announced in April 1977, the unions dampened militancy to avoid delegitimizing the leftist parties as the 1978 legislative elections approached. The CGT and the CFDT met with the owners and various ministers to argue against the job reductions. Protests were measured and well disciplined. In anticipation of a Left victory in the national elections, expectations mounted but mobilization was stillborn. The Left saw nationalization as a panacea for steel, and the popularity of that message should not be underestimated.[33]

Given the severity of the steel crisis, however, firm-level decisions did not await the outcome of the French political calendar. Furthermore, local union organizers were less able than federation leaders to divorce themselves from the immediate material concerns of displaced employees. The unions reacted to the 1977 Plan Acier through a combination of planned militancy and strategic action designed to highlight the plight of the steelworkers. The CFDT-FGM apparently contemplated and rejected an unlimited occupation of the steel plants.[34] The CGT and CFDT organized rallies in Metz, Thionville, and Paris. A *ville morte* (dead city) was organized in Thionville in which sympathetic shops closed their doors. Since neither the CGT nor the CFDT yet believed in the seriousness of the crisis, neither wanted to approve (*cautionner*) the job reductions. Both unions therefore refused to sign the social plan that specified how reductions would take place. They argued instead that nationalization of the industry by a leftist government would provide more generous social provisions.

As the 1978 legislative elections approached, the Left's steamroller lost momentum. Interparty rivalry exploded in September 1977, and the Common Program alliance broke down, thereby disorganizing the Left. The Left's defeat in legislative elections, in turn, altered strategic calculations for labor. The dashed expectations of March 1978 forced the confederations and federations to rethink their approach toward popular mobilization and political leverage. Rather than influencing labor policy by capturing key political offices and nationalizing the steel industry, the unions had to proceed without direct control over the state's powers of intervention. Instead, labor combined a hypermilitancy to get the state's attention with concrete bargaining to forward policy recommendations that would ease the pain of industrial adjustment. As happened so often before, though, interunion accord produced high levels of mobilization only on local labor policy issues. When central union leaders tried to use popular protest for larger political purposes, workers' enthusiasm for militant action declined.

Rumors of further cutbacks after the March legislative elections aggravated a widely reported sense of injustice.[35] The additional 20,000 redundancies threatened to destroy the leading role of the traditional steel regions. The importance of steelmaking to these areas heightened tensions in economic activities outside the steel industry, since so many other jobs depended on the industry. These cutbacks took place within a deteriorating national labor market in which a newly reelected Center-Right government supported investment rather than redistribution.

After Usinor and Sacilor formally announced their redundancies in December 1978, a variety of different and frequently dramatic forms of social violence rocked Lorraine, the Nord, and even Paris. While a demonstration in Metz might not be the first item on the evening news, the occupation of the Bank of France by FO militants could not be neglected. CFDT militants occupied regional television offices in Metz and broadcast appeals for solidarity. Autoroutes were barricaded. The entire town of Longwy was blocked. The CGT occupied police headquarters. A few zealous workers dumped tons of scrap metal and blocked the doors of the employers' federation. Unionists and non-unionists alike perpetrated these actions. They were seldom conceived through regular decision making in the union locals, although the latter usually claimed responsibility to head off the prosecution of individuals. The federations heard about them afterward.

These worker actions transcended typical modes of protest, such as demonstrations, sit-ins, and occupations. The aim was to damage property to dramatize the plight of the industry and its work force. Such violence was not new to the steel regions. Workers had undertaken equally dramatic actions in the past against employers. However, the 1978–1979 period could not be rivaled for the intensity of actions. One author chronicled 183 actions in the Longwy area alone.[36]

The spontaneous gestures enjoyed a media success: for six months the national press covered steel politics almost daily. A common perception among the participants was that union-led demonstrations would not be sufficient either to reverse the redundancies or rally popular opposition against the Plan de Sauvetage.[37] Militants saw the federations as impotent in the face of government action. This was partly a result of the shift in power in the labor movement. Spontaneous actions designed to dramatize industrial conflict characterized the behavior of local CFDT militants. In contrast, the CGT preferred more controlled forms of mobilization.

The unions both rode this crest of violence and sought to direct it. The CGT mounted a pirate radio station that called itself "Lorraine Coeur d'Acier" (Lorraine the heart of steel) and used it to mobilize broad segments of the population. Even Gaullists and Trotskyites received air time both to express outrage at the industrial policy decisions and to share their grief. Along with its CFDT counterpart—"Radio-S.O.S. Emploi"—it facilitated a broader social mobilization and cross-class opposition to the redundancies. It permitted the articulation of regional shock over the economic dislocation.

This cross-class and regional focus was a unique element in the 1978–1979 mobilization. Local merchants in Longwy and Denain shut their doors in *villes mortes* to protest the job reductions. Public servants used vacation time to join marches. Clergymen joined the demonstrations and spoke out from the pulpit. More than at any time in the history of the postwar steel politics, this conflict brought together groups of people who ordinarily did not support each other.

The geographic concentration of steel plants raised the stakes of mobilization. Until the construction of Fos, 90 percent of crude steel was produced in Lorraine and the Nord, with over 50 percent in Lorraine alone (see table 7.1, in the next chapter). Steel played a critical part in local economies. In towns such as Longwy or Denain, 80 percent of the economy was tied directly or indirectly to steel employment, which therefore assumed a broad social significance.

The 1978 redundancies ended the traditional steelworker family pattern whereby jobs were handed down from father to son. The isolated acts of violence reflected this grief. Mobilization tended to target the (dis)investment choices made by the companies. Even when the unions tried to regain control of the mobilization, they maintained a focus on labor market issues.

The mobilization became so decentralized that the unions sought to regain control. According to the CGT, "Lorraine Coeur d'Acier" lost its union flavor and was closed after the mobilization died down.[38] The CFDT actions against its union locals in Dunkerque and Longwy (to be discussed later) suggested a similar organizational response in which the unions sought to exert social control.

The unions calculated that violence could not be sustained and that some amount of demobilization was necessary to secure any advantage from the disturbances. The unions also feared that too many workers would accept the state's immediate offer—an exit bonus (*prime de départ*) of FF 50,000. For foreign workers it came to considerably more, since they could also cash out their social security contributions. This promised to undercut the unions' bargaining position, since it would defuse the mobilization which the unions were attempting to control.

Mobilization dwindled when centralized control was tinged with national political overtones. The CGT had proposed a march on Paris for March 23, 1979, and it called on other unions to join them for an interunion demonstration of power. Yet the CGT organized the Paris demonstration in ways that resembled its *syndicat unique* tactics of the 1940s. The Fédération des Travailleurs de la Métallurgie (FTM) left the preparation to the confederation. The Communists wanted to use the show of force to begin preparing for the elections to the European Parliament in June 1979. The theme changed from anger at redundancies to an anti-German, anti–European Community platform. The CFDT and FO pulled out of preparations and ordered militants not to participate. The acquiescence of all but a few locals (Longwy and Dunkerque in particular) broke the union front created at the local level. What had started out as local control of labor market mobilization fizzled into centralized control of political mobilization.

Thus, mobilization appeared unfocused. Yet it was part of the struggle for control within the steel industry. The changes taking place in industry governance had concrete consequences for redundant steel employees. Local unions initially sought to take advantage of regional anger by channeling and then directing it. Mobilization could serve as both a bargaining chip with the state and as a resource for interunion struggles.

From Mobilization to Propositional Unionism

Dwindling mobilization undercut attempts to use widespread protests to influence the adjustment process from the outside. By 1978, the unions sought to replace confrontational tactics with concrete negotiating positions designed to influence adjustment in steel by joining the inner circle of industrial decision makers. A generation of confrontation encouraged oppositional demands—"hard" positions intended to stretch management as far as it would go while simultaneously making a statement about class exploitation. Propositional unionism sought instead to reorient industrial development and to extend the collective bargaining from labor market issues to product market strategy.[39] Implementing the new strategy created organizational confusion for the labor movement. Propositional unionism increased central control and alienated the rank and file, for calculating the trade-offs of adjustment policy had not been part of oppositional unionism. Because the CFDT dealt with these conflicts more effectively than the CGT or the FO, it developed into the union best placed to negotiate over the terms of industrial decline.

The labor negotiations after 1976 took place entirely at the national level. The 1977 Convention de Protection Sociale covered both the Nord and Lorraine, although Solmer's plant at Fos and other, smaller plants elsewhere in France also agreed to adhere to the provisions. Its scope extended wider than the 1967 agreements. The age measures covered workers at all plants signing the accord. The uniform application of early retirement permitted transfers to other plants of the same company or different companies within the industry.

The increasingly national scope of negotiation increased incentives for concrete bargaining. The CGT and the CFDT agreed about many aspects of labor market policy, especially employment retention, but they differed over fundamental questions of production policy. In particular, the CFDT was more willing than the CGT to consider plant closings and employment reductions. The CFDT was thus torn between maximizing labor leverage to preserve union solidarity or departing from the CGT to embrace more "negotiable" positions.

The "Memorandum of the FTM-CGT on the Steel Industry" portrayed the steel industry as a technologically sophisticated industry that was being allowed to wither away by the actions of the state and the greed of employers.[40] It argued that the Plan de Sauvetage would bring incoherent investments and disinvestment that would contribute to inefficiencies and even future shutdowns. The "Memorandum" complained that the government was more interested in realizing short-term profits than long-term competitive capacity. It suggested that demand forecasts were incorrectly generalized from short-term swings in the market: the forecasts neither included public policies that would stimulate demand in France nor accounted for growth of foreign demand in the Third World. To attain the former goal, the task was to balance supply and demand for steel products. As a country with a relatively low per capita consumption of steel products, France should invest more heavily in economic infrastructure—roads, bridges, schools, and hospitals—to bolster domestic demand for steel. The companies themselves needed to invest both backward and forward: backward to incorporate the Lorraine iron mines in the planning process and forward into higher value-added metalworking activities. The "Memorandum" suggested various ways to improve the quality of working life and to preserve jobs.

The "Memorandum" was not as much a departure from previous strategic positions as their intellectualization. Its propositions did not reverse previous demands. For instance, it suggested shortening the work week, hiring a fifth weekly work team (for continuous operations), and abolishing temporary labor contracting to stabilize employment, but these were not new ideas. It also argued that Lorraine should remain the center of French steel production. The conflation of production levels with national *grandeur* recalled the earlier campaign for 35 MT. The "Memorandum" envisioned plant-by-plant investment levels to maintain production levels as high as possible and to reach 34 MT by 1983. (This would have meant an increase of more than 50 percent above 1978 production levels.)

The CGT's propositions were not new to the industry, as both the CFDT and FO had conceptualized the industry in similar terms ten to fifteen years earlier. The CGT, however, combined global perspectives with an unflinching defense of the steel regions: "Any definitive closure of a production unit, any falling back to limited market niches, any destruction of the historical gains of a region, any layoff of workers is a weakening of the entire steel industry."[41] Given the apparent unlikelihood of a Left government in 1978, the document was primarily intended as a set of negotiating positions with the industry and the state as well as mobilizational ammunition for class politics.

To considerably less fanfare, the CFDT-FGM issued "The Future of Steel: Propositions of the FGM-CFDT" in January 1979.[42] Much of the 1979 analysis was congruent with the CGT's "Memorandum." It contested the CSSF's complaints about price controls, arguing instead that the price differences did not explain the industry's decline and that the companies were more interested in short-term profit taking than long-term development. It disputed the industry's argument about salary costs: while Japan's wage bill was lower, Germany's was higher and both nations outperformed the French. The steel industry, according to the CFDT-FGM, had suffered from anarchic investment patterns, technological conservatism, and an absence of industrial "coordination." Government policy had contributed to the industry's demise by indulging the owners' management decisions and by confusing financial aid with industrial policy. It could not buy its way out of these problems; difficult industrial choices had to be made.[43] As in the CGT's "Memorandum," the CFDT's industrial propositions were combined with recommendations for a shorter work week, a fifth working team for continuous operations, quality of work life recommendations, and generalized early retirement at the age of fifty-five to provide jobs for younger employees.

The CFDT-FGM did not object to an industrial restructuring, nor did it reject decreases in levels of steel employment. It argued that the former should take place only in ways that strengthened the competitive capacity of the industry, thereby meeting national needs. The FGM would approve the latter only if alternate economic activity and employment were found for the steel basins. It proposed that the industry work from its strengths while simultaneously seeking to create a competitive advantage: trade could be managed (negotiations with the Italians and Spaniards over the importation of strip), while investment in research would meet the demand for new products from user industries. Investments in continuous casting and computer controls had to continue and accelerate. Eliminating product categories with weak demand—such as heavy sectionals or narrow strip—should take place cautiously so as to facilitate employment reconversion.

While the CGT-FTM insisted on keeping Lorraine the center of French steel production, the CFDT-FGM was willing to accept at least partially the government's policy of market niches that would affect disproportionately the production of sectionals in Lorraine. Likewise, although the CGT-FTM refused employment reductions, the CFDT-FGM sought to parlay them into new jobs.[44]

More important than interfederal differences, however, were tensions created within each federation by propositional unionism. Whereas union behavior is frequently initiated from below, promoting militants from the base to Paris guarantees a coordination of strategic perspectives between workplace union sections and the federation (or union locals and the confederation). The poverty of French trade unionism ironically ensures that paid officials remain in close touch with locals. Previous strategic changes in both the CGT-FTM and the CFDT-FGM had either resulted from gradual drift or had occurred over sufficient time to iron out differences. But the severity of the steel crisis of 1978–1979, together with swift strategic reorientation, created organizational tensions for both federations. As both the CGT and CFDT grappled with ways to respond to government initiatives, they created confusion within their own ranks.

The CFDT's positions in steel policy developed out of the confederation's explicit *recentrage* (recentering), an attempt to reformulate union activity away from the state and within those areas the union could directly influence. For both confederation and federation, *recentrage* remained an ambiguous rallying cry. It could be interpreted either as favoring local struggles over political action or as emphasising bread-and-butter collective bargaining without a strategic vision.[45] This confusion over confederal direction threw the federation's propositions in doubt.

The combination of *recentrage* and propositional unionism exacerbated tensions between CFDT federation leaders and local union officials. During the social turmoil created by the industry restructuring in 1978–1979, the federation expelled its local at Dunkerque and nearly cut loose its local in Longwy. The dispute in both cases involved the appropriate response to proposed redundancies. If *recentrage* took a reformist bent, it meant negotiating with the industry regardless of the CGT's positions. If it took a leftist direction, it would almost require the maintenance of trade union unity. Both locals had histories of such unity. CFDT-Dunkerque had already gained control of the works council and was known for its aggressive militancy. The local in Longwy had called on workers to vote for the Communist candidate against the Socialist candidate in the first round of the March 1978 National Assembly elections. Against the wishes of the federation, both locals marched with the CGT and the PCF in demonstrations in Paris in March 1979. Both locals refused to sign the July 1979 social plan governing the terms of employment reductions. Both locals had demanded closer coordination with the other unions. The two locals saw effective action only in interunion cooperation, while the federation feared that the PCF (with its political agenda) had assumed the leadership.[46] The federation claimed that both had been infiltrated by *autonomes*, extreme leftist, quasi-violent groups.[47]

The CGT-FTM was rocked by internal tensions similar to those affecting the CFDT. Proposition-force unionism was not an easy strategy to follow for those in the regions affected by the restructuring. The FTM never adopted it uniformly. Parts of the Communist Party wanted to repoliticize the struggles in steel, while others sought a redefinition of roles for the CGT. Oppositional unionism came almost as a reflex for many affected communities—especially Longwy and Denain. The dynamics of interunion competition influenced the confusion, as the CFDT-Longwy had assumed an oppositional position. The tension between oppositional and propositional strategies characterized CGT strategy through most of 1979—in other words, after the March on Paris in March 1979 and the European elections in June, both of which allegedly repoliticized the CGT.[48] The confederation even published a serious analysis of the steel industry in November 1979 that adopted most of the perspectives and proposals of the "Memorandum."[49]

The CGT's shift to propositional unionism increased its distance from the Communist Party.[50] Whether it took the form of political appeals or collective bargaining positions, oppositional unionism served the party's needs. It expressed the class rage from which the PCF worked. Even when the CGT ended its sectarian political activity in the 1950s and adopted a limited autonomy toward the party, the two organizations worked closely together, accepting a division of labor in pursuit of common goals. Propositional unionism extended that autonomy. Taken to an extreme, it even recognized that legitimate differences could separate the two organizations. Yet the risk was worth the possible pain, so argued some Communists in the CGT, because it would increase the efficacy of union action.

Both the CGT and the CFDT experimented with various strategies to minimize internal tensions and to maximize their inclusion and influence in negotiations over the regional dislocation that would result from massive redundancies. Union actors found themselves in a double bind, however. Oppositional tactics resonated most strongly among the local production employees in older facilities who were directly affected by industrial restructuring. Yet such protests mobilized workers for a losing battle because they were premised on saving uncompetitive jobs. In contrast, propositional unionism offered a way to increase the union voice in steel industry negotiations, but it centralized union strategy into the confederations, where debate was detached from the anguish of regional decline.

Of the three major steel unions, the CFDT was best able to combine mobilization with successful negotiations. The CGT attended most collective bargaining sessions, but its negotiators were caught by the abrasiveness of their programmatic demands, which to a new managerial class were not constructive. Throughout the postwar period, the CGT was unwilling to admit that production should be shifted out of old and into new plants or that employment levels at given sites should be decreased. In contrast, the CFDT, while sharing many of the objectives of the CGT, proved more capable of structuring bargaining positions amenable to employer negotiators.[51] The breakthrough on age measures in 1967, for instance, had a precedent in a firm-level accord negotiated a year earlier at the beleaguered specialty steel works at Pompey. There the CFDT had fought alone within the works council for early retirement as an alternative to layoffs.[52]

Because the CFDT had an image of being less "workerist" and more open to all categories of workers than the CGT, it caught the imagination of French intellectuals and the sympathy of the media.[53] Union archives show that state bureaucrats in ministerial cabinets had considerable sympathy for the union's positions.[54] Likewise, sympathizers for the CFDT-FGM could be found within the steel companies. For instance, the CFDT learned surreptitiously the contents of negotiations between the firms and the state in 1978–1979. The union had more information at its disposal and could sharpen its negotiating positions.

Finally, the CFDT could not have made such gains if it were divorced from popular protests. The contrast between the FO and the

CFDT is striking. While the FO had long signed contracts in the 1950s and 1960s that dealt mainly with wages, it could not gain widespread support from the negotiations in the 1960s and afterward. The CFDT developed a negotiating team in steel that eventually surpassed that of the FO. The CFDT never lost its popular base. Unlike the FO, the CFDT stood for labor unity and refused (until the 1980s) to denounce the local actions of CGT militants. The early role of Descamps had been critical to the development of the CFDT in Lorraine. The future secretary general of the confederation started a tradition in Lorraine of unity in action with the CGT.[55] The CFDT had militants who viewed the world in general and steel in particular in class terms. Thus, its militants gained respect even in the Longwy and Denain areas, both known for long histories of Communist and CGT hegemony.

The tragedy of the CGT was its incapacity to speak to new audiences. It correctly read the sentiments of its constituency, but the latter was shrinking. The rhetoric of *la casse* (destruction) played well with those displaced by adjustment. In spite of its energy and tenacity, the CGT gained the reputation of a union that refused industrial change. The CFDT's style of unionism combined the militancy of the CGT with flexible negotiating stands. The CFDT could thereby serve as a bargaining partner, and its signature was sought by the state even if its militants were blocking highways and fighting the police. The CFDT could return to its electoral constituency with negotiated benefits in hand.

Bittersweet Victories

The French steel unions fought rear-guard battles between the Liberation and the employment hemorrhage of the late 1970s. The CGT and CFDT/CFTC failed to achieve most of their strategic agenda, while the FO could congratulate itself for participation in the collective bargaining process that brought steelworkers important guarantees. Although the CGT and CFDT enjoyed some benefits from those collective bargaining concessions, they sought unsuccessfully a unionism that could negotiate the terms of industrial change and not simply the consequences of that change.

Propositional unionism was not possible in the steel industry, given the dominant institutional interaction of labor, capital, and the state. The very existence of competitive unions meant that working-class unity would occur only in historical moments and not on a consistent basis. The residue of paternalism prevented employers from sharing social control. The development of French steel capital resulted in the dispersion of capital and of the consequent regional bargaining structures. The government conspired with the companies until the employment crisis to exclude unions from decision making. It showed a marked preference for social stability, which it consistently tried to purchase with lucrative collective bargaining concessions for the companies. In this environment, the unions could have done little more. The cards were not in their hands.

The two successes of the union agenda—centralization of collective bargaining and nationalization of the industry—came from forces external to the labor movement. Collective bargaining became centralized more because the nature of the employment crisis demanded a national approach than because of union positions and strategies. A conservative state took control of the industry in 1978 through de facto nationalization and poorly disguised "participatory loans."

Locating the 1979 Convention Générale de Protection Sociale (CGPS) negotiations in Paris symbolized the centralization of employment bargaining within the industry. If the 1977 measures in the Convention de Protection Sociale foresaw interplant solutions to personnel reductions, few transfers had actually been implemented. The 1979 agreement offered greater incentives to workers to transfer. Moreover, greater pressure was placed on older workers in competitive plants to retire to make room for displaced workers. The *primes de départs* were also available to all workers to free up as many posts as possible for transfers.

Regional bargaining meant little to an industry that by the late 1970s had made production national in scope. Concentrating capital within the industry was a major goal of French public policy. By 1980 Usinor and Sacilor accounted for over 90 percent of steel production. Those two national champions produced steel in several regions and did not need to stabilize local labor markets. Moreover, the few workers who continued to be recruited did not necessarily come from these regions. The reorganization of work changed the type of employee in steel. Because job descriptions depended increasingly on computerization, workers had to be recruited away from employment elsewhere in France. The labor market for steel employment had become national.

The formal nationalization of the steel industry by the Socialists in 1982 marked the realization of a longstanding union demand, but it was not the panacea the unions had expected. Access to decision makers improved. Likewise, unions gained access to production and investment decision making, since members had seats on the boards of the two companies. Yet the industry continued to shrink, even as the CGT adopted as its rallying cry the most optimistic of the three hypotheses of the Judet Report—24 MT annual production. The constraints imposed by international competition limited production and employment even in a healthy industry. They left little room for maneuver in a declining one.

Nationalization of the industry destroyed even the semblance of regional bargaining units for the employers. The Groupement des Entreprises Sidérurgiques et Minières (GESIM) resulted from the fusion of Assimilor and the Chambre Syndicale de la Sidérurgie du Nord in May 1982. As the single interlocutor for the unions in collective bargaining sessions, the GESIM brought together the interests of Sacilor and Usinor and their subsidiaries.

The realization of union goals coincided with a general demobilization and with the full impact of international surplus capacity. National bargaining would not bring the expected higher wages in the midst of labor shedding. A nationalized company intent on reducing its work force and maintaining wage stability could be as ruthless as a private employer. As will be argued in chapter 7, the individualization of salaries and the plant-by-plant participation schemes rendered national bargaining obsolete soon after it was established.

Conclusion

State policy altered the governance of the steel sector in France. Still, many of the dynamics of adjustment remained in place. Throughout the postwar period, the state participated in forecasts and production choices, approved investments, and served as banker. Through price controls, it influenced the marketing of products. It established a labor policy based on buying support. Only by the late 1970s, however, did governments dare alter the balance of power within steel policy making. Yet the state's assumption of control of steel was expensive. Between 1978 and 1986 alone, the government spent an estimated 100 billion francs on the industry.[56]

The greater the intervention in the product market, the more union militancy targeted the state as the instigator of changes in the employment relation. For thirty years, unions had sought to place the employment relation in the industry on the public agenda. They sought nationalization in the belief that jobs would be more secure. As the state became deeply involved in the industry's financial well-being, those arguments assumed greater cogency. As jobs and regions were threatened by market forces, they assumed greater urgency. Nationalization of the steel industry in 1982 marked the culmination of its politicization in the postwar period. It was a Pyrrhic victory.

The steel unions were more successful in pursuing social than industrial issues. After the mid-1960s, the unions received employment protection, admittedly for a shrinking core of workers. Such success resulted partially from the dedication and perseverance of union negotiators. However, it became possible only because of the nature of militancy in the industry. The ability to achieve concessions was intimately linked to the power to invoke state intervention. Seen in this way, the weak class regulation and oppositional unionism resulted in a militancy that became functional to collective bargaining success.

As collective bargaining became centralized, union militancy remained decentralized, and a tension developed between union locals and the federation. This combination worked initially to labor's advantage. Federations were not held responsible for disciplining the rank and file, since attempts would have been futile. Consequently, a division of labor unfolded within (and between) the unions, separating bargaining from mobilization. Labor's interlocutors therefore faced two adversaries—union negotiators and militant workers. Both needed to be satisfied to ensure social stability.

Decentralized militancy, however, depended on factors that were increasingly irrelevant. Paternalism was less important as a target of mobilization by the 1970s. Work reorganization produced a more heterogeneous work force that was difficult for the unions to organize. Political cleavages among the confederations filtered down to irreconcilable differences at the federal level. With the shriveling of mobilization, the unions were left only to bargain with their limited organizational resources. As the tasks of competitiveness became increasingly stark, there was little to bargain over by the 1980s.

SEVEN

Competitive Success

A visit to a French steel mill in the early 1990s belies traditional images of steelmaking. These are not the smoky, dirty plants of previous eras. The shop floor is largely automated. Employees are highly trained in metallurgy or computer processing. They work in climate-controlled cabins sheltered from the heat and smell of metal transformation.

The dramatic turnaround of French steel had its source in policies developed in the mid-1980s and the Socialists' embrace of economic orthodoxy. Under the auspices of the capacity reductions ordered by the ECSC, the industry phased out the remaining old unproductive mills. It continued to purchase new plant and equipment. It decreased its labor force, insisted on a new relationship between manager and worker, and upgraded skill levels. It combined aggressive commercialism and cooperative labor practices. The French industry gained a competitive advantage by obtaining the flexibility necessary to compete in volatile markets.

Socialist "orthodoxy" succeeded where the previous conservative governments had failed for two reasons. First, the transformation of interests in the steel policy community enabled an unambiguous response to market turbulence. Second, integration of labor and industrial policy tools maximized productivity. This required flexible solutions to the problems of sales volatility and labor rigidity in the 1980s and 1990s. New patterns of workplace authority accompanied the reorganization of work. The decline of union opposition gave managers enough space to experiment with new plant-level social policies. Such experimentation found a mix of industrial and labor policy to promote industrial competitiveness.

Europe and the U-Turn on Steel Policy

The French steel industry had both prospered and struggled under the aegis of the ECSC. Until the 1970s, the Community was not a key actor in French steel policy. It protected the French market as long as producers respected posted prices. The prolongation of surplus capacity, however, created a new role for Brussels. It generated consensus among European producers both to protect the market better and to reduce overall capacity. European cooperation dovetailed with changing perspectives within the Left on industrial policy in general and steel policy in particular.

The Socialists came to power in 1981 with the pledge to "reconquer domestic markets"—a godsend to those industrial patriots in the steel industry who continued to measure tonnages. This was part of a larger attempt to "democratize" the French economy. Nationalization of eleven major industrial groups and the remainder of the banking sector would permit greater public control of the economy. Labor reforms would allow employees to have a greater voice in the process of industrial adjustment. Decentralization of public administration would bring government politics closer to the individual citizen. In the short term, stimulation of demand would revive the economy before the structural reforms took full effect.

That stimulation package quickly created its own set of problems—inflation, trade deficits, and pressure on the currency—which forced a policy choice on macroeconomic management. Economic pressure highlighted a series of political struggles within the government and the leftist parties that had consumed the previous two years: should economic policy be formulated within the constraints of the international political economy or should the Left give itself room for maneuver to implement its stimulus package and structural reforms by erecting barriers to the international economy? By March 1983, the Socialists had decided to replace economic voluntarism with caution on policy. After 1983, the Socialists substituted investment for redistribution, private initiative for nationalization, and Europe for domestic markets. "Modernization" replaced socialism as the dominant rhetoric.

The decision to pursue industrial policy within the constraints of the international (and especially European) market constituted a brazen change in political coalitions. Within the context of the strong Deutschemark and cautious financial management in Germany, remaining within the European Monetary System (EMS) would inevitably entail higher unemployment and downward pressure on wages. The Socialist government showed itself willing to withstand the protests of organized labor. The precise political calculations are less important for explaining steel policy, but the inability of organized labor to influence the policy agenda reflected the division and disorganization of the union movement.[1]

The evolution of the ECSC gave the French state the tools it needed to lower production costs and further reduce French steel capacity. European policy had evolved considerably since Ferry's failed attempt to encourage active intervention by the European Commission. The recession in the European steel market after 1974 had encouraged other steel industries to search for coordinated action. The Commission, however, delayed major actions until 1976–1977 at the instigation of the strongest producer, Germany. Three stages characterized European cooperation in addressing the problems raised by Ferry.

First, Henri Simonet, who held the industrial policy portfolio on the Commission, forged a set of policies designed to bolster the steel market. These included regional aids, better market forecasts, better protection against third countries, and recommended prices for steel products.[2] After 1976, the Community signed voluntary export restraint agreements with third countries to restrict the flow of cheaper steel onto the European market. (Third-country producers routinely sold below posted prices.) While the Germans applauded the trade policies, they viewed the industrial measures as too *dirigiste*. In contrast, Ferry found them insufficient because they did not immediately remedy low European prices that exacerbated the critical debt problems of the French firms.

Second, Etienne Davignon, Simonet's successor, took the crisis measures a step further in May 1977. He placed minimum prices on individual products—reinforcing bars, merchant bars, and hot-rolled sheet. The Commission imposed fines on deviant firms. Davignon had the support of the cartel of European steel trade associations—the European Confederation of Iron and Steel Industries (Eurofer).[3] However, individual companies took their cases to the European Court of Justice and fines were delayed. Moreover, by the time the cases were settled, the

European market had improved somewhat, and the fines were not large. In times of massive surplus capacity, the incentive was to cheat.

Third, the collapse of the steel market in 1979–1980 forced the Commission into even more drastic action (Davignon Plan II). Using its powers to implement Article 58, it imposed mandatory production quotas on four groups of flat products, and voluntary quotas were requested for other products in 1980.[4] The following year quotas were extended to cover 65 percent of Community steel products. The size of fines was increased to deter infringement. The Commission also strengthened the subsidy code by tying public assistance to capacity reductions. While capacity reductions were negotiated with all European firms, the special penalties for firms receiving public subsidies was a victory for the Germans, whose industry prospered largely outside the auspices of the government's financial support.

Cartels help those firms least able to cope with market changes. The European crisis cartel penalized the Germans and the Italian private sector, the small minimills in the Brescia region of northern Italy, the two strongest actors in the market. The Germans—focused largely on integrated production—were prevented from expanding their market position when other firms faltered. The Bresciani were clubbed into conformity with other members of the cartel, first by pressure from the Commission, then by Eurofer-engineered increases in the price of scrap. Thus, the subsidy code was small compensation to market winners for the political sacrifice of market position. The Commission-sponsored cartel gave French policy leaders the market protection to take the political moves necessary to restructure the industry.

The government of Pierre Mauroy decided to accept the capacity reductions decided upon in Brussels under the terms of the second Davignon Plan. To a government bent on economic orthodoxy, the continued losses of the two steel groups became intolerable. Usinor and Sacilor lost MF 15,500 in 1984—more than the combined losses of the world's top twenty steel companies!—and the government formulated the sixth plan in eighteen years to solve the problems of French steel.[5]

Adjustment and Profitability

To solve the problems in the French steel industry, the state needed to increase capacity use, lower costs, and focus production. In light of the massive losses of the steel companies and because other European producers were pursuing the same goals under the Davignon Plan II, French steel policy gained considerable urgency. Economic restructuring carried with it enormous challenges for the work force. Overall job losses were high, and the demand for highly skilled employees altered the social composition of the workplace. The changes affecting the industry required a labor policy capable of balancing both needs.

The 1984 steel plan once again cost Lorraine jobs. The MoI canceled the investments planned for Gandrange in 1982—a universal mill for multiple grades of sectionals. Likewise, a wire mill in Rombas abutting Gandrange shut down. Usinor-Neuves-Maisons closed its new oxygen converters (built in 1979) and replaced them with an electric furnace at a loss of 1,400 jobs. Outside Lorraine, the specialty steel plant at Ugine-Fos (built in 1975) stopped production. A total of 20,000 jobs became redundant. Another industry reorganization accompanied the new measures.

In a depressed market, the French needed more coordination of capacity utilization. The state created two subsidiaries of Sacilor (51 percent owner) and Usinor (49 percent) that grouped the production of sectionals (Unimétal) and construction-grade specialty steels (Ascométal).[6] The production of sheet and other flat products had been the only profitable activity for both companies. They remained temporarily with their respective companies.

The Unimétal merger covered a wide array of products that traditionally had problems.[7] With eighteen sites and production at 4.5 MT, it immediately became the largest European producer of long products. The plan for reviving these products—wire, pilings, joists, merchant bars, rounds for tubes—ended the "*tous azimuts* strategy" of selling an array of steel products. Production concentrated on areas where the new company was strong (such as wire products) and deemphasized those in which it could not compete (reinforcing bars).[8] It broke with the strategy of "reconquering domestic markets" and tried instead to enter other European markets (especially Germany and Italy). The Unimétal plan represented the first attempt to adapt French steel production potential to market conditions.

The most important production decision for Unimétal concerned the move toward electric arc furnaces. Steel capacity in electric furnaces would increase from 20 percent to 40 percent between 1983 and 1987. Three new minimills would be built in Valenciennes (under the name of the subsidiary Métalescaut), Neuves-Maisons, and Longwy. The SMN was slated to become a "midimill" with both electric and oxygen furnaces. The electric furnaces would give the company greater flexibility to confront demand fluctuations. The plan for Unimétal envisioned that oxygen steel capacity complement the minis. Although Gandrange was a relatively new plant, it was less productive than other new plants due to the high coking costs of Lorraine iron ore. The plan forecast a reduction of activity from 1.7 to 1.0 MT of production. Overall capacities declined by 12 percent.

The 1984 plan was part of longer trend that whittled away at Lorraine's dominance of steel production and employment (see table 7.1). While Sollac continued producing sheet, only isolated pockets of other steel production would exist. Of the 10,134 total redundancies in the Unimétal plan (more than half the announced job reductions), 9,368 would take place in Lorraine. After a brief interlude between 1981 and 1983, the Socialists had carried through the logic of Barre's reasoning in 1978: steel would be an industry like any other and should keep the same financial stringencies.

Table 7.1 French Crude Steel Production by Region, 1960–1986 (in 1,000 MT and percent)

	East		North		Other Regions	
	MT	%	MT	%	MT	%
1960	11.3	65.6	4.0	23.0	2.0	11.4
1970	14.0	58.8	7.1	30.1	2.6	11.1
1974	14.3	52.7	9.0	33.4	3.7	13.9
1978	9.8	42.9	7.6	33.3	5.4	23.8
1982	7.4	40.0	6.3	34.1	4.8	25.9
1986	5.3	29.7	7.0	39.0	5.6	31.3

Source: CSSF, *La sidérurgie française*, various years.

That logic implied a reduction of capacities and employment. Even after the dramatic changes in 1984, the industry continued its retrenchment. Another set of measures in June 1985 cost 1,350 jobs. Treating steel like any other industry meant reducing capacities that could be used elsewhere. Unimétal made a deal with the Luxembourg steel firm ARBED to exchange the production of joists and pilings for products that Unimétal would make. This arrangement increased capacity use for both firms but brought another 500 redundancies to Valenciennes.[9]

The problems of specialty steels had traditionally revolved around geographic dispersion and technological backwardness, especially for low-grade specialty steels. The 1982 mergers brought a consolidation.

The creation of Ascométal brought a much larger reorganization. The original plans in January 1985 called for work force reductions of 4,500 workers (out of 11,298), almost half of which were in Lorraine at the Société des Aciers Fins de l'Est (SAFE) and Pompey. The weak market in steel products forced additional reductions in capacity. The Dunes plant shut its oxygen converters, sourcing instead from Dunkerque converters, to increase overall utilization rates. Cooperation with Unimétal in semifinished products also meant better use of capacity. In a policy reversal, Ugine-Fos (less than ten years old), which had been slated for closure, remained open. Pompey shut down completely.[10] As with crude steel production, the manufacture of specialty steels required both technological change and managerial innovation.

The redundancies created by the 1984–1985 restructuring necessitated another set of "social" provisions that again precluded layoffs. The steel regions were designated "conversion poles" for which they became eligible for job-creating investment subsidies. As a symbol of its commitment to creating employment, the Socialists appointed Jacques Chérèque, the former deputy secretary general of the CFDT, as regional industrial coordinator for Lorraine.[11]

Change in government only strengthened the developing consensus around steel policy. The Right government elected in March 1986 appointed Jean Gandois—former board member of Sacilor and former CEO of Saint-Gobain—to investigate the industry. He recommended further redundancies (25,000) and capacity reductions (2 MT) and suggested a unified chain of command for Usinor and Sacilor. He claimed the two companies needed another infusion of public money (MF 18,000) to restore funds. These recommendations were implemented by Francis Mer, also a former CEO of Saint-Gobain who was chosen in September 1986 to head the merger that created Usinor-Sacilor.

French steel eventually returned to profitability. Usinor Acier (the sheet division of Usinor) and Sollac turned profits in 1986. The following year the production of sheet—from the sites at Dunkerque, Solmer-Fos, and Sollac-Florange—was consolidated in the subsidiary Sollac. The unified group Usinor-Sacilor was in the black for 1988 after fourteen consecutive years of losses. Profitability can be traced to four principal factors.

First, by the early 1990s, French steel had become a low-cost producer of increasingly higher value-added materials. This entailed a brutal reduction of capacity throughout the 1980s. It also meant investment in continuous casting and automated production such that the industry could point to a high level of technological competitiveness. It phased out open hearth and Thomas steel capacity. It had the European Community's highest percentage (82 percent) of crude steel production by continuous casting by 1985. By 1990 it maintained its lead with 95 percent of its steel continuously cast.[12] Between 1980 and 1988, as we see in table 7.2, French labor productivity jumped faster than for any of its rivals. Its experimentation with super alloys made France an important developer of specialty steels.

Table 7.2 Index of Crude Steel Output per Worker, 1988 (1980 = 100)

France	197.4
Germany	152.1
Italy	161.9
United States	164.1
Japan	117.8

Source: OECD, *The Iron and Steel Industry*, 1990.

Integrated production was centered at three sites—Fos, Dunkerque, and Sollac-Florange. A variety of minimills manufactured a narrow range of products for local markets and offered the company a new flexibility of production. By 1990, Usinor-Sacilor had the highest productivity levels in Europe.

Because of heavy investment in new production methods, the company pulled back from its attempt to diversify downstream and instead intensified its steel operations. This responded to the strongest criticism of the industry as it was losing money in the 1970s: the focus on steel made companies vulnerable to market fluctuations. Yet such vulnerability resulted from the product lines manufactured by these companies: crude steel producers had trouble serving as price leaders because rivals found it so easy to enter the market.

The development and production of sophisticated alloys, however, led to a different logic. Quality production brought a constant search by firm managers to improve products and entailed a commitment to basic research. It meant working closely with end users to meet their needs and deter them from substituting alternate materials. Quality production fostered a change in labor relations that encouraged contributions by employees to the production process.

Usinor-Sacilor's major commitment to downstream diversification during this period was consistent with a focus on quality steel products. The acquisition of Carnaud Basse-Indre in 1990 transformed Usinor-Sacilor into the world's third largest producer of tin plate. The purchase made sense, because Sollac already furnished 80 percent of Carnaud's raw materials. It was also part of a larger strategy to focus production on alternate materials.

Second, the industry was able to reduce employment faster than its competitors (see table 7.3). Throughout the 1980s, French steel was able to shed jobs. At first, this meant capacity reductions. Increasingly, however, process improvements brought savings in labor costs. Usinor-Sacilor reduced its work force faster than the major European producers after 1977. As chapter 8 shows, the employment reduction resulted from the state's capacity to mobilize the steel regions for "reconversion." The constant pressure for quality production and the new labor relations, however, ensured that job reductions did not have an overly disruptive effect on work organization.

Table 7.3 Employees in the Steel Industry in Selected Countries, 1965–1990
(per 1,000 employees)

	France	West Germany	Italy	United States	Japan
1965	158.8	255.6	69.7	584.5	321.9
1970	145.7	237.7	74.4	549.6	345.2
1974	155.7	230.6	93.6	525.0	323.8
1980	114.0	201.0	101.0	429.3	271.0
1984	87.1	156.5	81.7	267.4	264.8
1987	62.9	137.3	64.9	210.0	232.3
1990	48.1	127.0	57.0	194.5	275.4

Source: OECD, *The Iron and Steel Industry,* various years.
Note: Employees = process workers, technicians, office staff, workers in auxiliary services, and apprentices.

Third, Usinor-Sacilor finally achieved financial solvency. It did this through lower cost production, but the Right government gave this a boost in 1986 by "renationalizing" the two companies, dispossessing minority stock holders, and reconstituting the base capital in the merged company.[13] The financial moves provided a cover for the state to absorb losses and alleviate the debt. Yet the financial concerns of the group continued to require surveillance. Because of Mitterrand's 1988 campaign pledge neither to increase nor decrease the public sector, the company had been hamstrung in raising capital. That was largely alleviated by the 1991 purchase of a 20 percent stake in the company by the state-owned Crédit Lyonnais.

Fourth, the company moved aggressively to grab market share for steel and steel products. It marketed exports aggressively and purchased foreign steel assets to produce in other markets. Steel is not typically known as an industry with multinational corporations. Usinor-Sacilor changed that image by spinning a web of cross-border alliances through a series of takeovers and partnerships in the late 1980s and early 1990s. The purchase in 1989 of Saarstahl (through its subsidiary Dillingen Hütte) made the company the second largest German producer. Its takeover of Jones & Laughlin in 1990 gave Usinor-Sacilor an important presence in American markets to avoid import restrictions. That year the company became the world's second largest steelmaker.

Usinor-Sacilor has also focused on the internationalization of its commerce. It purchased distribution outlets in Germany, Britain, Italy, and the United States to gain access to customers, shorten delivery periods, tailor deliveries to client preferences, and thereby increase market share.[14] This marked a recognition that end users were seeking to lower inventories. French steel managers had learned effectively from both the Germans and Japanese.

By the 1990s, French steel had become a major competitive force in international steel markets. As we can see from table 7.4, it exported over half of its domestic production. The French market was no longer protected. Instead, the industry targeted niche markets where it could sell and relinquished those in which it was no longer competitive. Usinor-Sacilor no longer depended on state aid. Instead, it criticized those producers still dependent on state subsidies at the expiration of the Davignon Plan II (1986). The European solution to surplus capacity had given Usinor-Sacilor the protection to adjust—from both external competitors and internal critics. It also had a high social cost: between 1974 and 1992, French steel shed two-thirds of its work force. In the process, however, the company had become a major player in international markets.

The Shift to Quality Production

Technological change between 1945 and 1975 focused primarily on size of facilities and speed in the manufacture and rolling of steel. The decision to eschew minimills—and relinquish certain of these markets to the Italian producers—brought a very high concentration of modern equipment to French steel sites. In a period of expansion, the immense mobilization of capital could be recouped. During market stagnation, steel companies faced enormous pressure to maintain production, become more efficiently organized, and find market niches.

Process innovations and product quality improvements became the two keys to solving these problems. Every year between 1975 and 1988 the industry had lost money, so the companies faced enormous pressure to lower costs. Moreover, clients such as the automobile industry and oil exploration firms were demanding increasingly sophisticated specialty steels. In an attempt to compete with alternate materials, steel producers placed more emphasis on the manufacture of quality products.

The quality "revolution" affected the steel industry in full force after the late 1970s. In the four major centers of steel production in Usinor and Sacilor—Dunkerque, Fos, Sollac, and Gandrange—managers experimented with different ways to improve machine performance and the quality of output. By the late 1980s Usinor-Sacilor was the third largest consumer of computer equipment in France.[15] The technological upgrading of steel dramatically altered work organization.

Both Usinor and Sacilor made attempts to enhance production coordination. This resulted from a desire to wage a "war on waste."[16] Ma-

Table 7.4 Crude Steel Production, Consumption, and Trade in France, 1952–1990
(in million tons and percent)

	Production (MT)	Imports (MT)	Exports (MT)	Exports as % of Production	Apparent Consumption[a] (MT)	Imports as % of Consumption
1952	10.9	0.1	2.7	25.4	8.2	1.3
1960	17.3	3.7	6.1	35.3	14.9	24.9
1970	23.8	8.9	8.4	35.4	24.3	36.6
1974	27.0	9.9	11.4	42.2	25.5	38.7
1980	23.2	9.7	12.0	51.6	20.9	46.4
1984	18.8	8.0	10.6	56.3	16.2	49.4
1987	17.5	7.9	10.6	60.5	14.8	53.4
1990	19.0	9.0	10.4	54.7	17.6	51.1

Source: Eurostat, *Iron and Steel Yearbook*, various years.
Note: Imports and Exports in crude steel equivalents. Totals may not add correctly due to rounding.
a. Apparent consumption = production + imports–exports.

chine operators and engineers were sensitized to the costs of energy inefficiencies, waste of materials, breakdown of equipment, and so forth.[17] Automated controls in blast furnaces, for instance, dramatically lowered energy requirements. In a period of weak price levels, production could continue only by lowering costs or moderating future increases. System integration and technological upgrading became all the more imperative with the equipment shutdowns following from the mergers of the late 1970s. As prices rebounded in the late 1980s, however, cost-cutting norms had been well internalized within Usinor-Sacilor. The concentration of increasingly sophisticated technologies put a new premium on maintenance and repair. Where one production step depends on the next, timely maintenance avoids bottlenecks. Over 25 percent of production employees in Usinor-Sacilor were engaged in maintenance and repair by the late 1980s.[18] Greater technological sophistication demands more skills and wider distribution of skills for maintenance workers. Various schemes for flexible work organization (to be discussed) combined maintenance with production.[19] In the coking plants at Fos, for example, every production employee was required to have maintenance skills.[20]

The French steel industry also needed to compensate for a lack of expertise in the manufacture of steelmaking equipment. A longstanding criticism of French steel management has been its focus on engineering quality at the expense of profitability. Yet the major innovations in steel have been largely imported.[21] Rolling mills came from the United States, steel works from Austria, blast furnaces from Japan, and continuous casters from Germany and Japan. Steel producers needed to capture organizational successes to adjust for the distant relationship between steel production and steel equipment manufacturers.

Usinor-Sacilor had success with incremental improvements of imported equipment. The new emphasis on coordination also developed from the increasingly intricate organization of the production process itself. For instance, to avoid buildup of a "skin" of cooler steel that impeded the molten metal in continuous casters, Usinor-Dunkerque developed methods to insulate that flow in the early 1980s. Both the companies and the industry research group—Institut de Recherche de la Sidérurgie (IRSID)—spent considerable sums exploring technological and metallurgical innovations.[22] With the merger of Usinor and Sacilor, basic research was centralized and applied research distributed among individual plants. Particularly fruitful in this domain were experiments with ultralight steels for tinplating that compete favorably with plastics.[23] Likewise, Usinor-Sacilor in the late 1980s began joint research projects with regional universities and even with European competitors.

French producers also needed to take full advantage of available equipment. The most obvious way was to "intellectualize" work. The new equipment required increasingly higher levels of skill. The more automated the equipment, the more discretion needed by individual employees to interpret data or solve problems as they arose in the produc tion stage. The most recent phase of automation has engendered a dramatic increase and generalization of intellectual labor.[24]

The upgrading of skills that characterized the steel industry in the 1980s was part of a longer process, just described, but the changes were different in nature. So-called blue-collar workers were becoming a distinct minority on the shop floor. At Fos, for instance, Giusto Barisi found that in 1983 only a quarter of the employees were manual workers and that between 1978 and 1983 the numbers of semiskilled workers dropped by 45 percent.[25] Table 7.5 shows this process for the entire industry. Likewise, boundaries between employee categories were becoming blurred.

Table 7.5 French Steel Employment by Category, 1970–1985
(in percent)

	Process Workers	Administrators, Technicians, Supervisors, Managers
1970	75	25
1974	70	30
1976	67	33
1979	64	36
1982	60	40
1985	55	45

Source: CSSF, *La sidérurgie française*, various years.

The demand for intellectual labor has translated into a demand for training. The funds allocated to training more than doubled between 1982 and 1985.[26] Equally important, training changed in form and function. To link product quality to personnel training, managers needed a better understanding of the problems of upstream suppliers and the requirements of clients. For supervisory personnel, it also meant long hours of training in organizational forms to facilitate further learning.[27]

A clear example of this high-skill strategy was the Compétences program at Sollac-Florange. Begun in 1988, it sought to mobilize the entire plant toward efficiency and employee involvement by matching individual skills to production tasks, wider job definitions, more direct relations between employees and suppliers of parts and materials, and a commitment to training for each employee. The training budget doubled to over 6 percent of total salary costs by 1990.[28]

Industry upskilling has occurred within the context of work reorganization. The older occupational categories are being replaced with skill groupings based on wider skill definitions and more diffuse responsibilities. Multiskilling has increasingly become a goal sought by line managers. Work autonomy and the development of direct contacts with suppliers has dramatically decentralized production management.

Yet steel production needs complex coordination. For instance, problems in a reheating oven on a hot rolling mill might cause slabs to be heated incorrectly; slabs would be rolled at the wrong temperature; rollers would have to be accelerated to prevent cooling.[29] The solution to this dilemma of powerful and simultaneous decentralizing and centralizing tendencies has been systemic coordination of quasi-autonomous groups. Here the emphasis is on constant communication between interdependent parts of the firm.[30]

The increasingly intellectual nature of work has relegated most of the arduous, back-breaking work to the past. Exposure to dust, noise, and heat has decreased with the removal of blue-collar workers from the shop floor. Consequently, production accidents continued to decrease over the 1980s. It is subcontractors who face a much greater risk of injury.

The high-productivity workplace enabled Usinor-Sacilor to decrease working time throughout the 1980s. The firm was under some pressure from the Socialists in the early 1980s. By government decree, the work week was reduced to 39 hours in 1982. In May 1983 an agreement with the unions reduced the work week to 33.6 hours for continuous operations and added a fifth team for shift work. Another agreement in November 1983 reduced the working week for those in noncontinuous operations to between 36.5 and 38 hours.

The decrease in work time, however, was purchased with a slowdown in salary growth. Table 7.6 shows the relatively flat gains in wages over the 1980s. Usinor-Sacilor was able to pursue a policy of constant upskilling without paying significantly higher costs. By the late 1980s it kept salary increases among the lowest of the nationalized companies. This was partially due to the hemorrhage in employment—two-thirds of all steel jobs were lost between 1977 and 1990. Lower labor costs enabled the company to recoup its productivity gains.

Table 7.6 Hourly Labor Costs in Steel, 1981–1988
(in European Currency Units)

	France	Germany	Italy
1981	11.32	12.20	8.47
1984	14.80	16.07	12.14
1988	19.33	20.19	20.82
Nominal wages, $\frac{1988-1981}{1981}$	71.0	65.0	146.0
Consumer prices, $\frac{1988-1981}{1981}$	55.3	15.7	88.2
Annual real wage growth	2.3	5.2	5.9
Annual increase in steel industry productivity	9.0	5.3	6.7

Sources: Eurostat, *Coût de la Main d'Oeuvre*, various years; OECD, *Main Economic Indicators*, various years.
Note: Productivity = crude steel output per worker.

The changes in work organization that took place in the 1980s and early 1990s dramatically altered the relationship between employee and machine. Proto-mass production in French steel had become flexibilized. The poorly educated semiskilled worker had been mothballed. The goals of Sollac, according to its president, were "zero error, zero delay, zero inventory."[31] By the mid-1980s, the French steel industry had made the transition from "smokestack" to "high tech" status.

New Patterns of Workplace Authority

In the tumult of industrywide employment reductions, plant managers needed simultaneously to ease out older workers, ease in workers transferring from other plants, increase skills for all workers, and ensure quality production. Overcoming economic backwardness required a massive workplace mobilization. Some plants were more successful than others at inculcating a new ethic of professional responsibility, although eventually all plants experienced an extraordinary turnaround in authority relations.

From the late 1970s, however, management rallied around the themes of flexibility and quality. These, of course, were the buzzwords of business schools and an increasing percentage of companies in France. As with all business fads, they were not as dramatic as their promoters or detractors claimed. They also needed adaptation to the environment in which experimentation was taking place. The process of implementing a quality approach was profoundly political and it varied by firm, by plant, and even by shop.

The steel companies made a rapid but not instantaneous transition from paternalism to participative management. The paternalist project met its demise in the mid-1960s when the industry needed to go hat in hand to the state, when managers were recruited outside the old steel families and the trade association played a larger role in interest representation. State and society had changed, and the old methods of managing labor were no longer suitable. Finding an appropriate replacement, however, was no easy task. A set of firm-level bargained compromises with organized labor might have provided a replacement, as it did in other advanced industrial countries. Yet conditions were not opportune: the unions spoke with many voices, the late tenure of paternalism discouraged participation for either side, and the system of industrial relations impeded the development of joint solutions on the shop floor to managing labor. Collective bargaining developed after the mid-1960s, but not the micro-level union-management cooperation over work rules and work organization.

Thus, until the 1970s steel plants tried a number of experiments to solve the "labor problem." The companies reinforced the organizational capacities of personnel offices and renamed them "human resources" departments. Much study was given to the dehumanization of work and to autonomous work groups. Usinor created "social foremen" who were asked to explore workers' needs inside and outside the plant.[32] Still, there were important plant-by-plant variations. For instance, Usinor-Dunkerque tended to experiment with work groups more than Usinor-Denain, largely because it had greater absenteeism. The small marginal plants of the Nord and Lorraine sought to maintain employment at any cost and seem not to have experimented much at all. Training was increased for production workers in the decade from the Convention to the mid-1970s.[33] The national reforms in salary disbursement—the *mensualisation*—altered the system of remuneration from hourly to monthly for most categories of steel jobs. It regularized employment, formalized career advancement, and reduced the discretion of production managers.

Solmer-Fos pioneered in developing new methods of employee involvement. After a series of strikes in 1979 and the disorganization caused by redundancies, management launched several initiatives to stimulate participation in the full automation of the plant. In 1977 a "concertation commission" composed of mid-level managers and engineers had been created. It was asked two years later to put together a plan of "technological, economic, and social development of the plant."[34] The commission proposed and upper management accepted a plan by which first heads of departments met together and then mid-level managers met to decide short-term plans to increase performance.[35] Because these meetings involved few employees, the commission established suggestion groups (*groupes Eureka suggestions*) through which workers could recommend changes in operations that would be forwarded through supervisors to higher levels. Employees whose recommendations saved the most for the plant won bonuses and vacation travel.

Simultaneously, the commission established temporary study groups (*groupes de progrès*) on virtually every aspect of firm activity. They averaged five employees, crossed occupational and skill hierarchies, and involved more than 20 percent of the 6,900 employees at Fos. Groups defined problems in conjunction with management. Again, the motivation was a competition through which group members could receive points that could be cashed in through a catalogue of prizes. After identifying problems and solutions, the groups disbanded. Like the Eureka groups, however, these progress groups tended to involve employees higher in the chain of command.

In 1982 management launched a plantwide effort to engage all employees in searching for higher-quality production. After visiting Japanese plants and meeting with colleagues from other large companies, upper managers decided that "quality circles" would alleviate hierarchical distances. The newly created Division of Social Affairs trained middle managers and engineers whose task would then be to animate "voluntary" quality circles within individual shops or offices.[36] Again, the quality circles were small, averaging five to eight members. In contrast to the progress groups, they consisted of employees who worked together on a regular basis. They also tended to be more unstructured than the progress groups, characterized by brainstorming over ways to reduce error and prevent recurrences.[37] Topics were vetted by the animators, not management. Rewards were given for money-saving actions.

These participative management schemes were wildly successful in their first few years of operation. Management estimated that 90 percent of all recommendations had a positive impact on the plant.[38] Ironically in terms of the initial managerial objective to lower costs, the quality circles generated more suggestions aimed at working conditions than productivity improvements. This may have been their strength, however, since they seemed to have provided an important integrative and consensus-generating function. (Moreover, the progress groups had already facilitated the process of work reorganization.) Barisi found that the "Solmer system" of work organization was not as flexible as that found in other European plants.[39] The participation schemes partially compensated.

At Dunkerque the push for participative management was also successful. The plant had a reputation for fatal accidents during the 1970s. Management therefore began its campaign by inciting employees to work together more effectively. The quality circles were developed initially to reduce the accident rate. In 1982 there were as many as 250 quality circles at Dunkerque and they tended to concentrate in production—in steelmaking and blast furnaces. Partially because they did not involve payoffs, they decreased to 50 by 1988.[40] Yet the search for cost reduction continued through the 1980s. In 1985, management and the CFDT created shop and office councils (*conseils d'atelier et de bureau*) that combined the concerns of safety and competitiveness. This was part of an attempt to decentralize management to five subunits, each with its own budget, human relations department, and quality program.

Sollac-Florange started later to "modernize" its social relations. A similar commission of managers and engineers was assembled in the late 1970s, but it played a smaller role in pushing for new employment practices. Only in 1984 did it make a push to promote the image of Sollac. Already by 1982, however, management was pushing progress action groups (*groupes d'actions de progrès*) that functioned in a manner somewhere between that of the progress groups and quality circles of Solmer. They tended to be ad hoc and to involve proportionally few production workers. Most important, they separated proposals (made by the groups) from solutions (organized by management). They tended not to reorganize work but rather to seek incremental improvements in existing practices. There were no financial rewards for successful proposals. The push at Sollac's Lorraine plant did not really take place until the merger in 1987 of Usinor Aciers and Solmer with Sollac. In 1988 the company linked productivity improvements to profit sharing and decentralized the calculations.[41] Sollac's managers claimed to have achieved "total quality"—first at Fos and then later throughout the company—by the late 1980s.[42]

The new management techniques were partially sustained by slick advertising. At each plant, publications celebrated the particular achievements of groups or individuals. Brightly colored posters emphasized the contribution of each individual to the team effort. Contests and pay bonuses to improve productivity linked individual advancement to company performance. Corporate slogans—"Mr. Sollac is doing well, thank you!"—were used to motivate employees. A decade earlier managers could only dream of such internalization of enterprise culture.

Using integrative management techniques coincided with the development of expression groups mandated by Socialist reforms in industrial relations, the so-called Auroux Laws. This was fortuitous for management. The 1982 law required all companies to negotiate over expression groups. (Although an amorphous concept in 1982, such groups were intended to provide communication between workers and managers.) Since management in steel was already in the process of worker involvement, it was relatively easy to transform these assemblies into productivity-enhancing tools. The expression groups started first in production and later spread to offices. However, union delegates were not always kept abreast of meetings, the problems raised, or solutions decided upon by management. The groups tended to be led by managers. At Fos and Dunkerque, expression groups were quickly transformed into "expression groups for quality." At Sollac-Thionville, in contrast, management used expression groups to complement progress groups and to encourage process employees to communicate problems in production.[43] In general, the unions were marginalized from the early initiatives to enhance quality.

Some noteworthy exceptions existed. The CGT had tried initially to push for a productivity-employment deal that would link savings to jobs. It had some success at Usinor's cold rolling mill at Usinor-Montataire.[44] The CFDT was still trumpeting the Dunkerque shop councils three years after their creation and suggesting various mechanisms to coordinate actions with union locals.[45] At Sollac, a companywide agreement was reached whereby union delegates would be informed of discussions within expression groups, leadership of the groups would rotate, and time spent in meetings would be compensated by the company.[46] Nonetheless, the unions witnessed most of the changes from the outside. The participative management techniques were not to be objects of negotiations.

By the end of the 1980s, the quality circle fad had come and gone in most parts of French business. In the steel industry, however, it was no fad. Participative management was not a gimmick to induce employees to work harder, as it was elsewhere in the economy. Rather, it was part of much larger strategy of job enrichment and changing work organization in steel. It was a concerted attempt to ensure quality in process and product. In the 1980s, productivity increased faster in Usinor-Sacilor than in any other steel company in Europe.

By the early 1990s, active recruitment was taking place in the French steel industry.[47] This was partially to rejuvenate the age pyramid, but it was also a way to continue upskilling and to maintain the focus on "total quality." It could ease out those less qualified through the collective bargaining agreement and bring in those with advanced credentials. The glossy promotional materials generated by Usinor-Sacilor showed a high-tech company that would provide a solid future for its employees. The company saw itself in the forefront of managerial mobilization to market highly skilled products. Participative management techniques enabled Usinor-Sacilor to mobilize skills and motivate a work force.

The Collapse of Traditional Unionism

The withering away of established patterns of labor organization both facilitated and reinforced the rebirth of the French industry. Integrative management was bolstered by the dwindling union presence in the workplace. The job losses hit process workers the hardest, and these were workers organized by the CGT and, to a lesser extent, the CFDT. The unions had difficulty mobilizing the new categories of workers. The older discourse about the bosses rang hollow to college-trained workers employed in a company that was becoming a European leader in materials development.

Labor demobilization had its origins in the dynamics of intra-and interunion relations. The federations regained control over strategic orientations and reestablished consensual politics in both organizations.

Part of that control involved staking out positions at odds with the other federation. A residue of propositional unionism remained in both camps. Both federations continued to publish careful industry analyses throughout the 1980s. Both federations advised the Socialist government from 1981 to 1984 on preferred directions for industrial development. Both federations negotiated earnestly the reduction of working time in 1982 and the extension of the CGPS in 1984 and 1987. Yet the CGT-FTM described the world more in class and oppositional terms than in the 1978–1979 period.[48] The CFDT-FGM was more willing after 1979 to negotiate with the industry and collaborate with the FO in works councils and during industrywide negotiations. Moreover, the CGT-FTM and the CFDT-FGM had little more than active disdain for the other and rarely formulated joint positions. Union cooperation by 1990 was less apparent than at any time after the war.

Relying on tried and true recipes meant reverting to old reflexes. The CFDT concluded that it could not afford to become "an office of economic studies."[49] The industrial sphere was no longer its "domain." Industrial propositions, it claimed, even tended to have a demobilizing effect. According to this line of reasoning, propositions were effective only when they responded to the demands at the base. And when this happened it was usually in the form of opposition strategies. The FGM felt that it needed to be more active in collective bargaining, where it could achieve tangible benefits for its members and electorate. It would not pick a "good fight" and go down swinging. Rather, it promised to refuse "lost battles," even if such positions attracted support among CFDT militants! Its actions should not belong to a long-term political project—a CFDT criticism of the CGT and PCF. It would be practical and less abstract. By the mid-1980s, the CFDT was defining itself almost as much in negative as positive terms.

By the mid-1980s, the CFDT-FGM—at its Thirty-ninth Congress in 1984 the metalworkers merged with the miners to become the Fédération Générale de la Métallurgie et des Mines, or FGMM—suffered from strategic uncertainty. The federation sought to initiate a resurgence of collective bargaining—a reformist variant of *recentrage*. For the federation (and also the regional union in Lorraine, located in Hagondange) the issues became negotiating replacement jobs in the steel region and guaranteeing benefits to displaced steelworkers. The locals in Lorraine, however, saw the continued drain on employment, the dashed hopes for a Socialist presidency, and the need for trade union unity. The 1984 employment reductions that resulted from the Unimétal and Ascométal restructuring plans in 1984–1985 angered the militants, who rarely missed an opportunity to remind the government of François Mitterrand's campaign promises in 1980–1981—no job reductions in the industry. The result of these different perspectives between federation and base was to demoralize the steelworkers, since no effective opposition to the government's plans came from above.[50]

CGT-FTM mobilization appeared no more successful. The return to oppositional unionism foreclosed alliances with other unions.[51] Yet oppositional unionism after 1979 appeared overly opportunistic, since by 1981 the CGT was changing its perspective once again. The three years of Communist participation in the Socialist government led CGT militants to dampen their demands in the industry, although oppositional unionism was not replaced by propositional unionism. Rather, their strategy was rooted in approval of the nationalizations and a willingness to negotiate with the government and industry. The FTM greeted the 1982 Plan Acier with enthusiasm, partially because it strove for high production levels. It used its seats on company boards to argue against production cutbacks.[52] Until 1984 the FTM contented itself with its new role as administrator and friendly critic from the inside. As with the CFDT-FGM, the CGT-FTM was willing to be patient while the new management defined objectives in the newly nationalized companies. Like the CFDT-FGM, it was seduced by newfound access to decision makers.

The further cutbacks in the now nationalized steel industry constituted almost a *coup de grâce* to the steel regions and steelworkers. The policy reversal of a leftist government leading the charge in restructuring the industry was particularly demoralizing. After the Socialist U-turn in March 1983, budgetary austerity replaced reflationary macroeconomic policy. The approval of 30,000 additional redundancies and the cancellation of investment projects in March 1984 signaled that the Left would depart from its national voluntarism and work within a European plan to reduce surplus capacity.

A series of demonstrations—including a march of 150,000 persons in Paris in April 1984—greeted the government's announcement. The unions sought centralized control, but coordination among the unions spun out in centrifugal fashion. The CGT became more vocal in opposition, especially in denunciations of the Davignon Plan, but it was reluctant to move into oppositional unionism. The PCF itself emitted mixed signals. Four members remained in the government, but Secretary General Georges Marchais marched in solidarity with the steelworkers. The CFDT had moved away from uncritical support for the Socialists, but *recentrage* had prioritized collective bargaining over labor mobilization.

The 1984 mobilization demonstrated the changing sociological profile of the affected work force. The severity of the downsizing threatened managerial as well as blue-collar jobs. The Paris march brought together production workers and office staff as well as engineers and managers. Like its counterpart five years earlier, the mobilization also included shopkeepers, civil servants, and retired workers. Once again, union militants pursued guerrilla actions. Highways and train stations were blockaded. The Socialist Party offices in Longwy were sacked.

These hard actions, however, did not spark the cross-class opposition of 1978–1979. Both Lorraine and the Nord had experienced twenty years of employment threats, and people were growing weary. More important, hard actions tended to mobilize semiskilled production workers, large numbers of whom had already left the industry by 1984. Therefore, such spectacular displays of despair had almost a counterproductive impact in 1984: they demonstrated existing class cleavages.

Once the march was over, however, those employees who were better positioned to test the labor market embraced the new policy of retraining.[53] This individualist response tended to contribute to more working-class demobilization. The unions were not in a position to continue forging unity. The tensions between the two confederations filtered down into the federations and locals. Aside from the Paris demonstration, cooperation in organizing mobilization was virtually nonexistent.

The difficulties of mobilizing in 1984 were part of a more general problem in the French labor movement. Mobilization has not been easy among steelworkers or elsewhere in the political economy. Cycles of conflict have been followed by periods of relative stability, although the absence of class regulation contributes to a certain underlying level of industrial action. Labor mobilization in 1948, 1967, and 1978–1979 tapped into this sense of injustice. The steel industry is heavily concentrated geographically, which transformed employment reductions into regional threats. The legacy of paternalism contributed to suspicions of betrayal.[54]

Oppositional unionism confronted this injustice. Yet it needed to be combined with concrete demands—security of income and employment. When the CGT and CFDT pursued labor-market demands and refrained from party-political strategies, they began to articulate the injustice. Local control of this labor market mobilization meant that political issues were deemphasized. Union cooperation at the local level guaranteed successful mobilization. This success explains the FO's loss of popularity between the mid-1960s and early 1980s. Organizational and political tensions eventually sparked a disintegration of working-class unity.

Moreover, industrial change made oppositional unionism obsolete. For large categories of workers, steelmaking was becoming a high-tech job. At the very least, increasing numbers of technicians and engineers could not expect their sons and grandsons to follow in their footsteps. Class boundaries had become blurred. The class rhetoric of the CGT appealed to fewer employees, and the negotiating activism of the CFDT nurtured an instrumentalism in the work force to use the union to get the best possible deal—and then leave.

By the mid-1980s, militants in the CGT-FTM and CFDT-FGMM were exhausted and confused. They could look with pride to the history of struggles in which they and their fathers had been involved, yet the industry was shrinking and their children were leaving the steel regions. Most important, the labor movement itself was exploding in the steel industry. Unions were fighting each other more than management. When protests broke out, union leaders were in no position to pilot them. The explosion of union rivalry provoked an institutional discontinuity that left the two unions reminiscing about the past rather than anticipating the future. Simultaneously, the companies were actively mobilizing the work force.

Conclusion

The pattern of institutional interaction between steel companies and state agencies evolved considerably after 1945. Market decline created crisis conditions that demanded a reevaluation of the distribution of power within the steel policy community. Although the companies tried desperately to retain managerial autonomy, they had to relinquish it to their trade association to stave off insolvency. This type of mesocorporatism was effective as a medium for negotiation between industry and the state, but it broke down in the midst of market hemorrhage. Adjustment brought a nationalization of interest representation.

Mere control of industry assets, however, did not dictate a particular set of industrial or labor policies. The variation of the 1980s could not have been larger: industrial strategy shifted from national to global terrain, while labor policy switched from job protection to market flexibility. By the late 1980s, labor policy facilitated industrial adjustment and did not serve as a goal in and of itself.

The turning point for the French steel industry was the decision to accept the terms of steel policy negotiated in Brussels. When the Socialists came to power in 1981, they forwarded a voluntaristic vision of adjustment. The 1984 steel plan recognized the constraints on French policy—European surplus capacity. This had to be negotiated away; for the Davignon Plan to succeed, the French had to participate. The decision to abide by a European solution to surplus capacity meant that older facilities closed and newer ones remained open. To increase productivity even further, workplace organization gained flexibility through new forms of employee participation. That flexibility helped to gut the traditional working class while simultaneously providing the conditions for profitability. When the plan succeeded in reducing surplus capacity, Usinor-Sacilor rewrote the rules for international competition. It acted like other large companies and used its financial position to forge cross-border alliances to protect market share in individual national markets.

Along with aggressive marketplace behavior came a new vision of labor market representation. This high-technology industry needed a highly skilled, flexible work force. Modernization of the 1980s succeeded because the mobilization of skills and corporate sloganeering transformed the workplace into a highly participative arena. The combination of sophisticated communications and a demand for highly trained employees made the new authority relations functional to the new work organization. That new workplace, in turn, reduced costs, increased quality, and used new technology to full advantage. By moving directly from a paternalist legacy to sophisticated worker participation schemes, it skipped a stage of labor relations that most countries experienced in the heyday of mass production. The same state that presided over an industry as a bulwark of stable industrial employment spearheaded the new industrial relations.

French modernizers triumphed forty years after they set out to reform the steel industry. The vision of postwar modernizers had finally been achieved—a unification of industrial and labor policy. Because modernization had its costs, however, it could not be imposed from above. It had to be negotiated, and unions participated in negotiating the terms of employment reductions. The development of an informal company covenant changed the character of labor politics. The next chapter explores how the unions were drawn into this process.

EIGHT

Converting Interests in Steel

Competitive success required changes within the French adjustment process itself. Company attempts at self-governance foundered on their incapacity to translate new investments into competitive products. Old sites remained active and unproductive workers were kept on payrolls, imposing a drag on balance sheets. Industry self-governance meant the exclusion of labor, but that feature was whittled away by a state nervous about the social consequences of inevitable redundancies. Breaking the stalemate required linking product market and labor market policy to change the balance of power among the critical actors.

State actors never succeeded in insulating the policy process from the rough-and-tumble of democratic politics, as the modernizers had hoped. The intrusion of "politics" into policy making in steel was virtually a constant after the mid-1960s. Every state intervention after the Convention became a test for the political regime. Various French governments dealt labor into the adjustment process to forestall social unrest. The threat of mobilization seemed to impel state actions in 1967, 1971, 1977, 1978, and even 1984. The state became the "objective ally" of organized labor—regardless of the party coalition in power. Yet this was not an alliance of equals: labor gained only symbolic representation and a considerable administrative burden in the policy process.

This chapter explores the politics of labor market adjustment in the French steel regions. It examines the passive policy tools used to encourage employee exit from the industry (see chapter 2). While they were effective in their immediate goal of reducing the work force, they tended to confuse firm-level personnel management, and they were extremely expensive. The attempts at retraining and relocating displaced workers were only partially successful because new tools were used with old purposes—paying individuals to leave. Consequently, policy leaders developed a set of economic "conversion" measures that would lower costs and increase the quality of human capital. The state complemented these more active labor market policies with regional development in its efforts to reindustrialize the steel regions in the 1980s, although the projects had a mixed success. Active labor market policy and regional development were complemented by formal labor inclusion. By the mid-1980s, the labor unions became entangled in a web of consultative positions to oversee the implementation of labor policy. The new administrative roles for labor changed the character of mobilization, which in turn offered new behavioral incentives.

The Political Consequence of State-Led Adjustment in Steel

The company-level integration of investment and labor policy fulfilled one of the modernizers' dreams. Combining the economic with the social captured efficiency gains, facilitated planning, and even showed some potential for extending democratic governance such that employees gained a voice in workplace decision making. State intervention had the overall effect of altering interest articulation.

The modernizers had also sought to insulate policy from social interests, however. The organizational logic of selective industrial policy in France almost demanded that unions—especially oppositional unions—be removed from the policy-making process. According to one senior civil servant in the mid-1960s,

> Neither the businessmen nor civil servants want to discuss their plans with union people (with rare exceptions) because they know that (1) most union people aren't really qualified to participate, and (2) most of them have short-term political interests which tempt them to leak the results of committee deliberations as one means to maintain or advance their interests with the rank and file.[1]

Oppositional unionism did not inspire trust from either firm owners or state officials.

This justification for excluding labor grew weaker after 1966. Business-state relations showed that these actors could also have "short-term political interests." By the late 1970s, proposition-force unionism brought trained economists into union decision making, "qualifying them to participate." One civil servant even admitted that the CGT's analyses in his tenure at the MoI (between 1983 and 1985) were frequently of a higher caliber than those of the industry itself, although he claimed that their quality dropped considerably after 1984.[2] Moreover, the union concerns with consequences of industrial policy on the labor market became more legitimate with the threat of massive redundancies and demands for quality production.

The more the state intervened, the more it politicized the steel industry. Public pronouncements of future success only nourished disappointment when results were not obtained. Both unions and local notables talked constantly of "promises not kept." Union publications denounced industry bailouts, measures taken on the "backs of workers." Such business-state collaboration only fueled the arguments of the CGT and CFDT for nationalization.

As industrial policy in steel became politicized, electoral fortunes seemed to ride on its success. This was as true for the Right as for the Left. After the 1965 presidential election, the electoral difference between the two sides of a polarized political community narrowed considerably, and public outcry caused by industrial dislocation had the potential to decide critical seats in the National Assembly. The political calendar after the mid-1960s dictated that a test of electoral strength would take place almost annually.[3] That constant national political jousting directly affected industrial policy. The 1977 Plan Acier was timed for public announcement after the municipal elections of that year. The decision to implement another restructuring plan, after the 1977 plan proved insufficient, was delayed until after the 1978 legislative elections, a contest most observers thought would be close. The Socialists delayed their 1984 restructuring plan until after the 1983 municipal elections so as to avoid the electoral consequences.

The steel trauma created cleavages within French political parties. Local deputies did not refrain from breaking parliamentary discipline to criticize redundancies if they could gain electoral mileage. Representing Nancy for the governing majority, Jean-Jacques Servan-Schreiber blasted the decision to invest in Fos, calling it a "super-Concorde junior that endangers all of French industry and our potential in the Common Market."[4] He later spoke out against the Plan de Sauvetage on the pirate radio "Lorraine Coeur d'Acier" set up by the CGT. Four Socialist deputies in the Nord quit their party in protest over the 1984 Plan Acier.

Before 1981, the Left used the decline of Lorraine as an example of capitalist mismanagement. The nationalization of the industry figured prominently in the Common Program. The Communists denounced the social and industrial policies of the industry as undemocratic and unpatriotic. Once the Left was in power, Mitterrand announced to a Longwy audience in 1979, it would stop the redundancies and rebuild the steel facilities. For both the PS and the PCF, this industry had to be nationalized to save the regions, protect jobs, and maintain steel production as part of the national heritage. The future of the steel industry sat prominently on the electoral platforms of both the Right and the Left.

State investments in the success of industrial policy and the electoral concerns of various political regimes combined to reinforce *for the unions* the logic of state-led adjustment. The leader of the CGT-FTM, André Sainjon, expressed this perspective after the announcement of the Plan de Sauvetage:

> It is perfectly clear that the State will hold control of the three steel groups. Therefore, it will be directly responsible for the industrial and social measures that will be taken. It has become the principal interlocutor.[5]

Consequently, union action took the form of a state focus that was sharper than that had historically characterized the French labor movement. Not only did mobilization target the political regime, but also negotiating strategies depended on state intervention.[6]

Because of the failure to link industrial and labor policies, however, the latter became ad hoc attempts to appease the unions over the terms of industrial change. Numerous measures were used to reduce steelworkers by whatever means possible: buying them off, retiring them early, and shifting them into different occupations. Such passive policy tools controlled entry and exit, but they could not change the quality of human capital.

In short, centralized bargaining together with decentralized militancy created uncertainty in industrial relations, which the state could only address in a very clumsy manner. The relative absence of class regulation increased the stakes for the interventionist state. Given the separation of militancy from bargaining, however, the costs of adjustment would consist of whatever was necessary to buy social peace.

The Consequences of Passive Labor Market Policy

The policy tools of the French state made limited forms of labor adjustment possible in the steel industry. Technological change and market competition forced most steel companies in the advanced countries to reduce employees after the early 1960s. Labor policy required not only reducing the work force but also promoting higher skill bases.

Product market adjustment in the French steel industry required a massive reduction in the work force (see table 7.3, above). Between 1965 and 1990, employment dropped by roughly 70 percent, and jobs decreased from 156,000 to under 50,000. Part of the job loss resulted from capacity reductions. Each of the four major restructurings—in 1966, 1977, 1979, and 1984—required significant job cuts. Employment losses also came from technological innovations in steelmaking. The generalization of continuous casters by the early 1980s, for instance, eliminated a number of different job categories. The introduction of computer controls for material flows in blast furnaces and steel converters had a similar effect a decade earlier. New technologies have altered skill requirements and complicated the tasks of labor adjustment.

Personnel management has always been difficult in the steel industry. Even before the period of mass unemployment, labor adjustment was a difficult task. Workers tended to have little education—even those with training had machine-specific skills—and less economic mobility. That difficulty was recognized by the ECSC when it established a social fund for redundant workers, although the fund financed only a small percentage of labor market programs.

A dilemma facing French policy makers and employers has been how to ease workers out while simultaneously defusing social tensions. Various schemes have been negotiated to streamline employment, yet they had to be implemented in ways that did not disrupt production. They have resulted in the anomalous consequence of loss of jobs without unemployment, at least among steelworkers. Despite the shrinkage of the industry, very few workers have been laid off.

Consistent with mass production techniques, labor policy focused initially on sheer reduction of steelworkers. The vast majority of surplus labor in the four years of the Convention was accommodated by early retirements and retraining and not by transfers. Retraining within the company schools brought some order to the hiring process. Fewer workers would be hired in the first place, and the work force was compressed from the bottom as well as from the top of the age categories. The Convention also compensated for the loss of skills caused by early retirement.

The Convention Sociale of July 1967 took as its point of departure the employment and transfer provisions of the *table ronde*.[7] The key provisions of the agreement included early retirement at sixty (at 90 percent of previous gross salary, based on a forty-hour week, and somewhat higher levels for the lowest paid workers), transfer provisions, and training possibilities.[8] Early retirement was extended beyond individual plants that needed to reduce employment to apply to the entire steel region (Lorraine and the Nord). Transfers for redundant workers were intended both within and between plants. However, because of internal transfers, workers received a high percentage of their previous salary only for the first six months (the previous firm would compensate at 80 percent the difference between the old and new salary). External transfers entailed the same benefits but brought lump-sum severance payments and required the approval of the worker. Any worker opting for a training program was guaranteed employment elsewhere in the industry.[9] A tripartite committee monitored the agreement.

The Lorraine accord was supplemented and improved in 1971 with the "Wendel-Sidélor conversion" into Sacilor. (Not surprisingly, the extension followed on a series of strikes at Wendel-Sidélor's plant at Knutange that was scheduled for closure.) Two changes deepened the original agreement. First, its duration was extended for an indeterminate period. This transformed labor adjustment from an ad hoc into a permanent policy of the industry. Second, new transfer provisions guaranteed equivalent skilling levels, and workers were promised 100 percent of previous salary for the following six months. This provision made the transfer to Fos more enticing, since the new plant changed skill qualifications. In its amended form, the Convention Sociale remained in force through 1976.

The economic downturn in 1974–1975 confounded political leaders who wielded traditional macroeconomic remedies—budget deficits and transfer payments—to stimulate consumption. The Chirac government believed that unemployment would be temporary while governments primed the pump to pull the economy back into the virtuous circle of growth. It also used a variety of tools to put the brake on job losses. Industrial policy during this period turned toward explicit employment supports facilitated by the creation of the Comité Interministériel pour l'Aménagement des Structures Industrielles (CIASI) and the Comités Départementaux d'Examen des Financements des Entreprises (CODEFI) in 1974 within the Industry Ministry.[10] Both the CIASI and the

CODEFI injected funds into the private sector to prevent bankruptcies. These measures were not unique to the steel industry, but they did indicate a state voluntarism that unemployment could be stemmed via government decree.

The Chirac government used several labor market mechanisms to keep workers employed. Legislation in 1975 forced employers to seek the approval of departmental labor inspectors before calling for a layoff exceeding ten employees.[11] This administrative control of redundancies (under the legal category of *motif économique* which had to be certified by the departmental employment services—DDE—and the labor inspector) tended to translate into state approval, except in the steel industry. Because of the industry's troubled history and political volatility, the prime minister forbade all layoffs between 1975 and 1977. The prohibition on layoffs meshed with the optimism of public policy and electoral uncertainty.

Also during 1975, the government reformed the system of short-time work (*chômage partiel*) so firms could receive additional state aid. This system enabled employees to remain juridically attached to the firm and either work reduced hours or stop working altogether for a limited period of time. Lost time was remunerated at 50 percent of the wage rate, with provisions to increase up to 80 percent under special circumstances. Under the 1975 legislation, employers could sign a FNE *convention* with the DDE to finance part of their share of compensation for the reduced hours.[12] Steel companies took advantage of this system to reduce work hours by 10 percent in 1975.[13]

The Barre government after 1976 rejected Keynesianism for a mildly deflationary policy stressing firm-level competitiveness and a strong franc. A different orientation in the governing coalition only marginally changed the orientation of labor market policy, however. It picked up an earlier idea of facilitating the departure of foreign workers. In 1977 it offered them lump-sum bonuses of FF 10,000 to return to their native countries—largely Algeria, Morocco, and Tunisia. Both Usinor and Sacilor used this option and reduced workers in these three ethnic categories at twice the rate of native French steelworkers between 1974 and the end of 1977.[14]

The expansion of early retirement was a more effective tool to whittle down employment rolls. This was facilitated by an experimental change (to last five years) in the Garantie de Ressources pour les Salariés Licenciés (GRL) in 1977, by which employees did not need to be laid off to tap into benefits. Financed initially by the UNEDIC, this GRL was used as an early retirement mechanism. Firms were encouraged to take advantage of this coverage to open up space for labor market entrants. Under the Garantie de Ressources pour les Demissions (GRD), they could resign from their employment and still receive the same compensation. The 1977 Convention de Protection Sociale (CPS) used this tool to compensate redundant workers over fifty-nine years old.[15]

The CPS lowered the thresholds for the age measures and improved the transfer provisions. Early retirement was formalized at age sixty, while early leaves (*cession anticipée d'activité*—CAA) were encouraged for those workers older than fifty-four years, thereby effectively lowering the early retirement age. Workers age sixty and older received 70 percent of their previous gross salary until age sixty-five.[16] This marked a slight setback for the unions from provisions achieved in 1967 when workers between sixty and sixty-five received 90 percent of their previous gross salary.[17] (Given the electoral fever, only the FO signed the 1977 agreement.) Those workers fifty-nine years and older received 90 percent of their previous gross for one year until they qualified for state-financed early retirement schemes. Those who left at fifty-six years, eight months, received 90 percent for the first year and 70 percent until age sixty. Workers at age fifty-four could qualify under certain circumstances (depending on management's willingness to release them) to receive 75 percent of their previous salary until the age of fifty-six years, eight months, at which time the relevant provisions would apply.

Continued reliance on age provisions promised to flatten the age pyramid, which could, in turn, distort skill levels. The employers consequently negotiated discretion in executing early retirements. That discretion created later problems when workers were unable to leave when they desired. The 1977 agreement made early retirement mandatory in all steel plants. Nonetheless, the age measures covered only half of the necessary redundancies. The rest came from "natural departures" (death, disability, turnover) and transfers. Those workers transferring to other plants received most of the same advantages as in 1967. In addition, more emphasis was placed on transferring to subsidiaries, other companies, or even nonsteel employment in the vicinity. The tripartite monitoring committees facilitated job searches.

The flattening of the age pyramid also made future age measures more difficult since, in the absence of new hiring, employment became increasingly concentrated in a smaller number of age cohorts. The Plan

Acier of 1977 was insufficient to stem the companies' losses almost immediately after its announcement.[18] The social provisions in the 1979 CGPS made even more radical choices to get workers off payrolls.

In 1979, the Barre government had reoriented a supplemental unemployment benefits program (AS-FNE) to cover labor shedding in the steel industry.[19] Starting with the social plan in steel in 1979, it lowered early retirement age (CAA) for workers over fifty-six years and two months, with exceptions, to fifty-five years. These bridging payments—funded at 70 percent of previous gross salary—carried redundant workers until they qualified for the UNEDIC early retirement programs. Both the composition of this funding and the compensation to the individual worker were later subject to revision. The compensation was decreased to 65 percent in 1982. At first, the UNEDIC paid 58 percent and the AS-FNE paid 12 percent of the individual's gross salary. After 1982 the UNEDIC part decreased to 53 percent. To lower costs to the state, however, part of the FNE contribution came from the beneficiaries' severance pay and financing from the firm—that 12 percent minus the severance pay. Once again highlighting the discretionary powers of the FNE, exceptions to this co-financing could be granted to the company if it were in financial difficulties. This exception became the norm, at least in the early years of this program.

The most blatant use of passive labor market measures was the *prime de départ* of FF 50,000 to renounce all claims to employment in the steel industry. The payoff was accepted by 6,000 workers.[20] Even this was insufficient, however, and the industry still needed to lower employment levels by another 15,000, and the older age cohorts were shrinking. Less than 4,400 workers older than fifty-four years remained in the industry.[21]

The 1979 CGPS further lowered early retirement age and loosened the remaining restrictions on transfers out of the firm. It created still another category of age measures. Those workers between fifty and fifty-six could qualify for a temporary leaves (*dispense d'activité*—DA) during which workers remained juridically attached to the firm and received 79 percent of their former salary. In the unlikely event of an increase in demand, workers could be called back. Since few observers actually foresaw such a circumstance, however, juridical attachment to the firm kept this method of reducing employment from being generalized elsewhere. The DA were again co-financed by UNEDIC and FNE.[22]

The 1979 agreement dramatically improved the conditions for

transfers to other jobs. Workers had the right to refuse two offers made by the personnel office. In the case of transfers to lower-paid positions, workers kept their seniority and their "coefficient" (pegged to the *minima*). They received compensation at 80 percent between old and new salaries (in other words, the bonuses that bumped up *minima* to real salaries) for two years. Transfers to other steel firms brought equivalent seniority levels but only promises to attempt maintenance of coefficients and real salaries. If promises could not be delivered, the worker received lump-sum payments. Transfers outside the steel industry brought lump-sum payments if salary levels were not matched.

The CGPS sought increasingly comprehensive solutions to the employment problem. The possibilities for transfers externalized the internal labor market. While the companies wanted to maintain a skill mix, they also tried matching their workers' skills to the needs of competitors or to those companies in the regions. Consequently, human relations offices became private employment offices. Not surprisingly, all five unions signed the 1979 CGPS.[23]

The transition to a Left government did not reorient labor market policy. In spite of its rhetoric and its initial enthusiasm, the Left discovered that the most effective means to contain unemployment in the short term was passive employment measures. Outside of steel, the number of workers age sixty to sixty-five years leaving the labor market increased by 60 percent nationwide between 1981 and 1984. The number of CAA's (fifty-five- to sixty-year-olds) leveraged with the AS-FNE increased from 11,551 in 1981 (almost exclusively in steel) to 99,296 in 1984, a 750 percent increase![24]

A state report criticized the state and the steel companies for laxity in the application of the CGPS, however. A new arrangement was needed in 1984, Jean Rosenwald argued, in part because the companies hired 19,000 new workers between 1979 and 1983. Although 64,000 workers left the industry during this period, only 45,000 total jobs were eliminated. According to Rosenwald, early retirement at age fifty cost the state FF 500,000 per person, while departure at fifty-five cost FF 216,000. The report suggested using all the policy tools foreseen in the original agreement, including stronger training measures. It served as the intellectual foundation for the 1984 CGPS (to be discussed later).[25]

The continuing necessity to shed labor put stress on personnel decision making. Distorted age pyramids drew criticism from new and old managers in the nationalized companies by the early 1980s. The Socialists followed through on their campaign promise and encouraged GESIM to negotiate a fifth team for continuous operations, thereby reducing the work week (without loss of pay) for over half the production workers to thirty-three hours, thirty-six minutes. The dual forces of technological innovation and market competition made further work force reductions necessary. The CGPS, whose original duration was to have only been two years, was extended on an annual basis into 1984 until other methods could be negotiated.

The twenty-year employment crisis in the steel industry produced a variety of measures to soften the effects of industrial change for the work force. By the mid-1980s, workers were using the early retirement plans to leave the industry at age fifty. Transfers within companies and to outside work, either directly or under the auspices of training provisions, had led to virtual employment entitlements. Although supplemental aids came from the social budget of the ECSC, the French state provided the majority of the social cushion.[26]

The purchase of employee exit was extremely expensive for French taxpayers. Several estimates were trumpeted in the mid-1980s—the most common of which claimed over FF 100 billion had been spent between 1966 and 1984—to show the extent to which the state had "rescued" the industry. These figures undoubtedly underestimated the full cost of total intervention, since the age measures and training provisions were paid directly by the FNE. Whatever the real amount, however, steelworkers received generous compensation for relinquishing employment. As the president of the Court of Accounts stated in 1984, the state had feared social unrest for the previous twenty years and had consistently sought to buy social peace at a high price.[27]

Even costlier, however, were the effects on personnel management. The rigid form of employment reduction created havoc with age pyramids and skill levels. In an era when production organization put a premium on employee flexibility, passive labor market tools worked according to strict seniority. This demand for flexibility pushed the companies and the state into alternate forms of labor market instruments.

The Politics of Conversion

The costs of labor market exit encouraged policy makers to find alternative solutions. Passive tools also had the disadvantage of becoming politically unacceptable. The dramatic increase of nonemployed fifty- to sixty-year-olds bore testimony to the human cost of redundancy: throughout the 1980s newspaper accounts followed retired men, their families, and the loss of working life. Allowing companies to lay off workers, of course, was even more visible. The GRD and GRL programs were discontinued in 1984 to bolster UNEDIC. Public policy therefore needed to bridge the gap between active and passive measures.

The unions and GESIM negotiated a new CGPS in July 1984 to facilitate the employment cuts presaged in the government's April 1984 Plan Acier.[28] Although the social provisions were presented by the Socialist government as part of a comprehensive redevelopment of Lorraine and the Nord, they were more a continuation of existing policy to diffuse the impact of concentrated job loss. Lorraine and parts of the Nord were designated "conversion poles," therefore eligible for state subsidies. The strategy offered incentives to small and medium-sized businesses either to migrate to steel areas or to increase investments at preexisting facilities.

Conversion became the cornerstone of the 1984 CGPS.[29] With all the appropriate modifiers, conversion became the new buzzword in French social policy, evoking images of mechanical permutation and religious proselytizing. Although the precise meaning of the term remained open to debate, it always connoted measures that facilitated worker adjustment.[30]

The new labor market policy combined the traditional tool of early retirement with a "conversion leave" (*congé de conversion*) for others. The leaves suspended (but did not break) the work contract for a period up to two years. During this time, the individual received 70 percent of the previous wage and retraining in a new occupation for a period up to two years with promises of two job offers after the completion of training.[31] The FNE financed a third of the salary and half the training costs with the company paying the remainder. If the workers were not "converted," they reverted back to the company, were laid off, and received the standard unemployment package.

The 1984 CGPS further externalized internal labor markets. Redundant workers under fifty were eligible for a training-conversion programs (*contrat formation-conversion*). Workers who remained attached to the firm (DA) received 70 percent of previous gross salaries while they participated in a two-year retraining program for non–steel industry skills.[32] Each trainee was guaranteed two job offers in his or her new skill area. A later modification permitted workers to take the equivalent in lump-sum payments in exchange for renouncing all claims to steel employment. Thus, transfers outside the company and the industry had been formalized within the framework of retraining provisions. By 1986, over 5,000 workers at Sacilor were "externalized."[33]

Economic conversion in the steel industry had mixed results. The goal of encouraging exit from the companies was attained. Between 1984 and 1987, the industry shed almost 25,000 employees (28 percent of the work force) via the 1984 CGPS. Between 1987 and 1990, another 15,000 jobs were cut (23 percent) through the 1987 CGPS. Securing jobs for individual employees was more difficult.

The industry established outplacement facilities in each steel region—Lorraine, the Nord, the Centre, and the Mediterranean. Unimétal and Ascométal combined their operations in 1986 under the banner of Uni-Asco-Conversion. (Sacilor, Sollac, and Usinor maintained separate administrations.) Each company had operations on each site. For instance, Unimétal-Conversion had sections in Longwy, Gandrange, and Neuves-Maisons in Lorraine and in Valenciennes in the Nord. Each site unit reported directly to the plant manager. It assumed three functions—supervision of training programs, placement (Unimétal-Service), and management of workers who had been physically handicapped through occupational injury.

The social affairs and human resources department usually developed a list of redundant employees. Each worker had an individual interview—accompanied by a union official to act as counsel—with a member of a conversion unit to explore options. Applicants took a skills test to determine appropriate training level. They participated in various "trade fairs" to refine job preferences. The conversion services tapped the state employment services for available offerings. If training was warranted, the applicant and placement officials negotiated an appropriate program. Virtually every site and every company had innovations on this design. At Unimétal's plant in Gandrange, computer terminals had access to placement bulletin boards. Sollac advertised positions through the state-run telecommunications system (Minitel).

The training programs were run through several agencies. The departmental labor inspector worked with the reconversion unit to place individuals in existing programs in training centers located in the affected areas. The AFPA went to individual job sites to set up some training programs. The Ministry of Education opened specialized training centers for automated production, electronics, computer science, and management training in five different Lorraine cities.

A number of programs were developed for the difficult-to-train applicants. In Longwy, the AFPA established a special program that offered remedial courses in mathematics. It gave courses in reading and writing for the illiterate applicants, as well as courses in French language for foreign workers. More important, conversion officials developed individualized reinsertion programs—ranging from specialized drivers' education courses to physical fitness development to communication skills training—to complement the technical course of study.[34]

In spite of relatively generous provisions, only a plurality of redundant steelworkers selected the training measures.[35] Some chose the age measures. Still others cashed out their reconversion benefits voluntarily either to start a new business or buy an existing one. This correlated strongly with the level of previous training. Those employees who had skills upon which new skills could be developed tended to opt for the retraining programs. Those who lacked previous training, were old enough, and had sufficient seniority chose to exit the labor market. Those with highly marketable skills cashed out their benefits, taking their bonus to their new employment. Table 8.1 gives an indication of the diversity of possibilities. According to Jacques Chérèque, less than 5 percent of all those who sought to change skills by mid-1988 had failed in their endeavor.[36]

Lorraine was not the only steel region that succeeded in finding work for displaced employees. After the merger of Creusot-Loire's steel plants into Usinor in 1983, the former parent company went into bankruptcy the following year. Because the remaining operations concerned not just steel but other metalworking jobs, the reconversion operation was only partially covered by the CGPS. As with Lorraine, however, the Ministry of Labor threw its support enthusiastically behind the reconversion effort. Again, the results were similar to those in Lorraine: short-term success (over 90 percent of displaced employees found jobs).[37]

The costs of these measures was large. As mentioned, the Rosenwald report calculated that the age measures averaged FF 216,000–500,000 per redundant worker to bridge the gap to retirement. A later report put the average age measure at FF 1,000,000. Even the training programs had an average cost of FF 300,000.[38] Moreover, Usinor-Sacilor only bore 2.5 percent of total costs of early retirements.[39] Given the involvement of its training subsidiaries, the training usually cost the company more than the early retirements.[40]

The desire to rejuvenate the age pyramid led Usinor-Sacilor to try increasingly to shift the burden to age measures even by the late 1980s. In 1990, it announced plans to hire 1,000 school leavers a year to rebalance the age pyramid.[41] The technological changes demanded skills that the company would rather obtain by hiring young degree holders than through training programs. A virtual tug-of-war existed between a Ministry of Labor anxious to meet budgetary goals and a company trying to gain production flexibility.

Still, success must be measured in multiple ways. Jobs in the steel industry had historically brought high-paying stable employment. Those workers who cashed out their benefits frequently found themselves vulnerable to business cycles. One follow-up study of the former workers at the Pompey specialty steel plant found that one out of three who had started small businesses had failed. A majority (68 percent) had seen salary levels decrease. A high percentage (18 percent) was un-

Table 8.1 Unimétal and Ascométal Employment Reductions in Lorraine, 1984–1987

No. of Active Employees		
June 30, 1984	23,710	
May 31, 1987	11,006	
No. of Reductions	12,704	
Methods of Reducing Employment	**No.**	**(%)**
Age measures	4,399	(34.6)
Completed CFC:	3,451	(27.1)
Reclassified with job (12.0)		
Cash-out to start business (5.4)		
Salaried employment (9.7)		
Transfers elsewhere in steel	1,627	(12.8)
Return to home country	228	(1.8)
Special services for handicapped	194	(1.6)
CFC in progress:	1,061	(8.3)
"Hard cases" still unresolved (1.5)		
Uni-asco service	1,562	(12.3)
Turnover	182	(1.4)

Source: Convention Générale de Protection Sociale de la Sidérurgie du 24 Juillet 1984, *Bilan Lorrain de trois années de conversion* 1 (July 1987): 124–25.

Note: CFC = Training-conversion contract (*contract formation-conversation*).

employed five years after having initially found employment through a training-conversion contract (CFC).[42]

Conversion captured the fantasy of political leaders who saw the combination of training and leave as a way to lower the unemployment rate and prepare workers for new jobs. The Socialists pressured business and the unions to negotiate an accord to broaden the application of the measure throughout the economy. In the absence of agreement, the government passed legislation in 1985 that universalized the *congé*. The revamped *congé de conversion* offered redundant workers 65 percent of previous salary (in line with the benefits for unemployment compensation and early retirement) and a four-to-ten-month training period. Like the steel plan (and a similar plan in shipbuilding), participation was voluntary. Unlike its predecessor, however, the *congé* broke the work contract in order to incite workers to leave.[43] If converted workers did not find employment, they would again be eligible for the standard unemployment benefits negotiated by labor and management. The FNE would finance half their salary and training expenses.

The Right furthered the conversion nexus. When the government pushed through the law ending the administrative handling of layoffs in 1986, it encouraged business and labor to agree on new procedures to deal with redundancies. A collective bargaining accord in December 1986, adopted in legislation the same month, modified the existing arrangement on employment security and required a conversion convention be offered to redundant workers (under age fifty-six) in the case of all layoffs.[44] The *convention de conversion* lasted a maximum of five months during which redundant workers received 70 percent of previous salary (but without severance pay). If they still could not find work after the conversion period, they would receive the regular UNEDIC allocations with the proviso that two months of allocations be subtracted from the maximum benefit period.

The 1986 national accord/legislation improved upon the traditional measures of paying people not to work, but it was not an effective policy tool. Its training periods were insufficient. The training programs in place (as opposed to the plans in steel and shipbuilding) sought not to teach fundamentally new skills but to increase the possibilities for reentering the job market. Trainees learned basic skills, ways to present themselves to new employers, and job search strategies. While the information and communication aspects of this "training" were not negligible, they did not "convert" redundant workers.[45]

In spite of the fanfare given to conversion tools under the Fabius and Chirac governments, the measures remained vastly underused. Between 1985 and 1987 less than 15,000 workers used the *congés* and only one-third succeeded in their "conversion." The *conventions* attracted equally few participants in 1987—only 11,500 with a reconversion rate of 17 percent![46] Participation doubled in 1988. Weak participation rates partially reflected a restricted pool of eligible candidates. Age limitations as well as seniority restrictions (workers need two years in the previous job to qualify) excluded those most likely to apply, such as older and marginal workers. Employment opportunities have not always been well publicized.

Within the steel industry, workers preferred cashing out rather than taking advantage of training opportunities.[47] Because of the additional costs of training over cashing out, the Ministry of Labor had little incentive to convince them otherwise.[48] Unpublished data for the entire Usinor-Sacilor indicates that 75 percent of redundant employees under the terms of the 1984 CGPS (1985–1988) cashed out their benefits sometime during the course of their training-conversion contract. (A mix of training and cashing out was possible.) The use of individual conversion programs decreased further during the 1987 CGPS (1988–1991) when less than 5 percent of those in the conversion program were actually choosing to remain in the industry or to take advantage of the training provision.[49]

Between 1988 and 1993, Socialist governments played down conversion in steel and in the economy at large and tried instead to use limited-term tax incentives to hire new workers. The Ministry of Labor also urged employers to calculate more precisely their staffing needs. It added resources to the Ministry of Labor to advise companies on human resource management, in particular, with the introduction of new technologies in small businesses.[50] In general, French employment policy favored "insertion" into the economy over "conversion." If employment policy between 1984 and 1988 blurred the distinctions between labor market exit and what we might call labor market loyalty, governments after 1988 reversed that tendency. They staked out the loyalty terrain because they saw the needs of the individual and those of the firm as inseparable.[51] The ministry sought to generalize throughout the economy those participative management techniques discussed in chapter 7.

Economic conversion worked at an uneven rate. On the one hand, the companies sought to enact labor policy as quickly as possible to improve industrial productivity. On the other hand, training took months or years to succeed—that is, to result in a new permanent job for the individual employee—and creating jobs in areas affected by downsizing could take even longer. A clear tension existed between the two time horizons. At a national level, conversion, as practiced in the mid-1980s, encouraged companies to shift the burden of labor adjustment onto the state.

Regional Development to Solve the Employment Dilemma

The ultimate goal of conversion was job creation. At least two perspectives jockeyed for priority: should public policy bring workers to jobs, or vice versa? While the former implied developing the human capital represented by displaced workers, the latter involved regional development. Both strategies had inherent logics for social interests. Focusing on displaced steelworkers brought with it empathy for the costs of industrial change in which the state itself had participated. Moreover, aiding victims of industrial change brought the immediate political gratification of responding to social needs. Yet targeting displaced steelworkers to the exclusion of other workers affected by weak labor markets risked placing steelworkers on a pedestal, thereby creating a new labor aristocracy. In contrast, a focus on the region demonstrated a recognition of the multiplier effects of employment contractions. Since steel employment led to other jobs—for suppliers, subcontractors, shopkeepers, municipal workers, teachers—the sudden loss of tens of thousands of steel jobs brought a contraction of overall employment. Yet targeting the region would not guarantee jobs for those individuals most affected by the redundancies.

With the 1982 nationalization of Sacilor and Usinor, the Socialist government forced the companies to create "industrialization" firms that would help fund the creation of new jobs in the affected regions. Usinor created three—Sodilor for Lorraine, Sodinor for the Nord, and Sodicar for the Ardennes affected by the redundancies in the former Creusot-Loire facilities. Sacilor, in turn, created Solodev for Lorraine, Somidev for the Fos area, and Socadev for the Normand conversion. These companies ranged in capitalization from MF 50–100 and offered loans to companies to start up or expand operations near steel sites.

The industry made significant claims about the success of these companies. Sodilor was boasting in 1983 that its loans had already led to the creation of 360 jobs in 1982, 127 of which were at the Thomson videocassette plant (which closed operations in 1987).[52] A year later, Usinor claimed that its industrialization companies had led to a total of 2,500 new jobs.[53] Company-sponsored reindustrialization continued into the late 1980s. Usinor-Sacilor announced in its 1988 annual report that it contributed to the creation of 3,000 new jobs in 1987 and hoped to increase that number to 3,000 by 1990.[54] It exceeded that number the following year with a contribution to the creation of 5,500 new jobs.[55]

The Court of Accounts was less enthusiastic. In a report issued in 1988, it depicted inefficiencies, lapses of administration, and laxity of oversight. In one case, Sodilor spent MF 12.9 on a telecommunications firm that promised 514 new jobs and only created 14. Sodinor spent MF 16.4 on a firm that went bankrupt before it employed anyone. The bankruptcy rate for firms subsidized by the conversion companies fluctuated between 15 percent for Sodinor and 35 percent for Sodilor. Although not leveling charges of corruption, the Court concluded that neither the state nor the conversion companies had the competence to create jobs. It determined that the efforts of the conversion companies between 1982 and 1985 had cost MF 604 for a total of 10,065 jobs, only 20 percent of which went to former steelworkers.[56] Still, at roughly FF 60,000 a job, the funding was consistent with other efforts.

In response to the national jobs hemorrhage and provoked by the impending employment reduction in the steel industry, the Socialist government in March 1984 established fifteen "conversion poles" in depressed areas that were eligible for employment creation measures. These conversion poles constituted attempts to smooth out the functioning of different policy communities. The DATAR (Délégation à l'Aménagement du Territoire et de l'Action Régionale) had been created in 1963 to promote regional development. It served as the main coordinator to set priorities and facilitate linkages among the various spending ministries. The decision to create an additional administrative layer via the conversion pole was an admission both that the DATAR had not been effective in its mission and that the employment hemorrhage had outstripped existing administration. The conversion poles brought together the ministerial field services and DATAR both to attract additional investment to distressed areas and to ease older workers out to make way for youths entering the labor market. The prefect presided over the efforts and reported to the prime minister over the progress of these poles.

The commitment to a jobs-based regional reindustrialization was further indicated by the appointment of Chérèque from the CFDT as regional coordinator for the reindustrialization of Lorraine, a special subcabinet position. His task was to monitor and coordinate efforts to bring jobs into regions suffering from the loss of steel jobs. The Chirac government reappointed Chérèque after the Center-Right won the legislative elections in 1986. Between 1988 and 1993 he served as minister-delegate for regional development and conversion, reporting to the minister of industry.

The attempt to bring jobs to workers was frequently trumpeted by those in political office. The Barre government announced in late 1978 that it attracted an investment from Peugeot for an assembly plant in Lorraine. Eventually, the deal fell through because of the costs of the employment subsidies, although a parts plant was created in the Department of Moselle that employed a few hundred workers. Before the renegotiation of the 1984 CGPS, then Minister of the Interior Fabius announced the impending creation (with state incentives) of 4,000 new jobs. A year later, Chérèque announced amid fanfare that the promise had been kept. The goal of employment policy in steel was both to spread out employment opportunities and to create new jobs.

Such claims must be handled with care. Between 1984 and 1986 the fifteen poles created 27,000 jobs nationwide. The cost per individual job, however, depended on what was included in the budgetary calculations. If it included only targeted regional aid to spur investment, then the per job cost was FF 47,000—significantly less than the conversion measures. If it included other funds from the Ministries of Industry and Public Works as well as other regional monies and subsidies from the European Commission, then the per job cost increased to FF 215,000.[57] Moreover, questions remained about the wisdom of expenditures and the durability of those jobs. It was also not known which project would have continued to generate jobs in the absence of funding, thereby creating perverse incentives. By the late 1980s, some companies appeared to be searching for investment or employment subsidies.[58]

As with the individual conversion programs for steelworkers, regional development efforts had mixed results in creating stable employment. A paradigmatic case of the difficulty of employment creation came with the transformation of the Sacilor plant in Hagondange into a theme park. The site had belonged to Sacilor via the merger with SMS in 1967 and had been part of the original UCPMI (see figure 3.2, above) which had resulted from seized German assets in 1919. Both the conversion companies and Chérèque's operation backed the development of a plan that would attract consumers (especially Germans) to the area to spend money and create further employment. The site was razed in 1986 to make way for a Smurf Park (Big-Bang Schtroumphs), and the theme park constituted a MF 720 investment. Run by a private company, Sorepark, it benefited from MF 50 regional aid through DATAR as well as the participation of key actors in the financial community (Crédit Lyonnais, Société Générale, Paribas, and Suez), Sacilor (via Sodilor), and even the government of Kuwait.[59] The company had hoped to employ as many as 800 employees, half of whom would be permanent and half would be seasonal. It found itself inundated with fifteen times that number of applications for employment for its May 1989 opening.[60] Yet already by July 1989 the company was facing complaints from the unions that agreements on working conditions were not being honored, and sporadic strikes broke out. By autumn 1989 the company was cutting jobs as the anticipated daily demand of 25,000 to 35,000 visitors never materialized.[61] Four years later it was struggling to stay out of bankruptcy.

In contrast to the Smurf Park, the reconversion efforts in the Longwy area were longer lasting. In July 1985 the Socialist government in conjunction with the governments of Belgium and Luxembourg and the European Commission designated the region around Longwy a redevelopment zone. The European Development Pole (PED) promised to target investments to take advantage of and reinforce existing skills, infrastructure, and available staffing. It sought to coordinate infrastructural improvements and development initiatives among the three countries. It included a business park that would straddle three borders as well as a common technology training center. The intent was to create 8,000 new jobs (5,500 in France) over ten years through the construction of an industrial district of small firms (with a population 300,000) that combined the flexibility of small size with the socialization of resources to compete successfully in neighboring markets.[62] The combination of resources would allow effective marketing. The PED benefited from aid given by the Commission's Regional Fund to supplement DATAR monies.

This cross-national experiment enjoyed some success. By 1989, the

PED had attracted over 3,000 total new jobs and 1,500 on the French side alone.[63] The investments originated from nine different countries. While the incremental increases in employment were relatively small—ranging from 10 to 250 employees—they seemed durable. Still, Belgium and Luxembourg attracted investment at a faster rate, and questions were raised in the early 1990s about adding supplementary tax advantages for locating in France.

The less publicized efforts to encourage investment and to increase training levels have probably been the most successful in attracting longer-term stable employment. Usinor-Sacilor seemed to have learned from these experiences in regional development. The unified steel company combined its reconversion efforts into one company each for Lorraine and the Nord. In Lorraine, Sodiest began targeting its funds to smaller enterprises with much more modest expectations. It developed an evaluative expertise and follow-up capabilities that enabled it to work with companies, assess investment needs, and channel funds to projects that had long-term viability. By 1989 the bankruptcy rate of assisted companies was approaching zero.[64]

The focus on targeted projects was consistent with labor market policy after 1988 and the Socialists' return to power. The governments of Michel Rocard, Edith Cresson, and Pierre Bérégovoy sought increasingly to target particular subpopulations—long term unemployed, youth, and so forth. The redevelopment targeting was simply another version: it picked investment projects likely to stand on their feet and provide jobs for the medium term. This was also part of a general policy to incite employers to calculate more precisely their staffing needs. The Ministry of Labor baptized this policy—measures to encourage long-term management of the labor force through improved forecasting, more concertation with employees, and greater training efforts—"negotiated modernization." Part public relations, part serious policy, it promised to promote consensus and reduce bottlenecks at the micro level.

Still, the attractiveness of "big ticket" projects could at times become irresistible. After the announcement of further employment reductions at Usinor-Sacilor in 1990 and 1991, DATAR and the Ministry of Industry arranged for investment subsidies to Renault to expand its automobile plant in Meurthe-et-Moselle and to the Korean firm Daewoo to build a television assembly plant in Moselle. Together they secured an increase of 1,000 new jobs.[65]

Governments in the early 1990s were torn between immediate labor market solutions and longer-term redevelopment projects. This was not an enviable choice. The former were politically expedient but expensive and frequently unsuccessful in creating long-term quality employment. The latter were less expensive and more durable but politically objectionable because of long time lags. Both solutions were combined to detraumatize the steel regions.

The Administration of Social Protection

Neither the state nor the steel industry solicited union involvement until the mid-1970s, except to quell industrial unrest. Afterward, centralized collective bargaining facilitated greater contact with governmental officials. Contact did not imply influence, however. These audiences seemed designed to inform the unions of decisions already made. By this late stage, the financial drain on the companies was so severe that policy makers were handicapped in their options, whatever their initial inclination. The norms of the steel system favored the infusion of investment funds into the industry. They did not favor participation.

The position of the CFDT here was interesting. As a union that talked participation combined with technocratic planning, it curried considerable intellectual favor with officials in the ministerial *cabinets* before 1981. CFDT-FGM archives indicate that some confidential information passed between *cabinet* and Federation.[66] Yet the content of that information was typically confined to advance notification of decisions already taken.

The first year of the Socialist government proved to be the exception to the rule. Ministers of Industry Pierre Dreyfus and then Jean-Pierre Chevènement maintained regular contact with the unions over the new industrial policy for steel. Union influence is difficult to discern, but the optimistic reading of the steel market was consonant with union aspirations. One official in the MoI claimed in 1985 that several hundred meetings with the unions had taken place in the previous four years. However, the government ran the risk of union disillusionment, for after 1982 it listened politely to union arguments and then proceeded to formulate its own policy. Jacques Chérèque, who one year later would be appointed to coordinate redevelopment in Lorraine, complained about the lack of meaningful participation:

> We are invited to tripartite ceremonies often after everything has already been decided. In administrative practice or in concrete

> cases, there has been no break with the customs of the right. The officials of our industrial federations often have a great deal of difficulty obtaining serious discussion with their corresponding Ministers.[67]

While Chérèque may have dramatized the relationship somewhat, he pointed to the dynamics of industrial policy making.

With the massive employment dislocation, policy makers decided to include the unions in the policy process. Consequently, the unions have played a variety of consultative roles. Union actors have sat on boards that executed the various social measures. They have participated on the regional industrialization committees. They advised the prefects on departmental employment issues. They counseled the subsidiaries set up by Usinor and Sacilor to find jobs for redundant workers. They co-administered the conversion programs. Permanent consultation was not new—unions sat on oversight committees in the Ministry of Labor, for instance. It simply mushroomed.

The Convention initiated these consultative positions in 1966. It required regular consultations with the unions over the success of the redundancy measures. These sessions resembled the meetings of the sectoral commission of the Plan: Ferry would hold court and inform the unions of decisions already taken by the state and the industry.[68] Any debate focused on measures to cope with redundant workers. After the expiration of the Convention, the CSSF continued to deflect attention from industry decision making.

Nationalization perpetuated this outsider status of labor in industrial policy making. Although union representatives sat on company boards after 1982—positions that meant important access to information—they did not participate in investment or employment decision making.[69] This had the paradoxical effect of tying worker representatives to policies with which they disagreed. Yet the formal constraints of their office pressured these worker board members to act in a "responsible" manner.

The consultative role was particularly painful for a union movement uncertain of its strategic orientation. Only the FO seemed satisfied with the possibilities held by these consultative bodies. It could use these new positions to demonstrate its administrative competence in a manner similar to its claims in elections for its other institutional positions—on social security boards and arbitration councils. Both the CGT and the CFDT vacillated between accepting and rejecting the consultative role.

On the one hand, they could either make marginal improvements in the *lives* of displaced workers or take credit for new jobs created. On the other hand, such participation confirmed their relegation to the "social" sphere. Particularly after 1979, it signified an admission that union influence failed to secure employment in the steel industry. Given these dilemmas, the CGT had the most difficulty adapting to consultation. The CFDT's *recentrage* brought an acceptance of the union's role in social matters.

The consultative positions formally outside the steel industry (regional redevelopment and regional employment commissions) carried additional dilemmas. The reindustrialization measures questioned the mobilizational strategies of the unions themselves. Attracting industry to the steel basins meant glossing over the public image of an angry work force that might discourage new investments.[70] After all, industry does not go out of its way to find militant workers! This was more an issue for local Left politicians and civil servants in regional development programs, since their participation was more direct in the process of attracting business to the steel region. Yet the unions were implicated both in the process and substance of regional development policy.[71] They too sought to attract employment for the region shaken by disinvestment. This has been a dilemma from which the CGT and CFDT have had difficulty escaping. Their only recourse was to participate on a *personal* level, thereby avoiding organizational responsibility.

Consultation overtaxed already lean union organizations. Union organizers tended to have multiple responsibilities.[72] The decline in employment—including the acceptance of departure bonuses or cash-out options—by some of the most active militants put additional responsibility on those remaining in the unions. Faced with additional administrative tasks, union locals have witnessed a decline in vitality over the last fifteen years.[73]

The consultative role did not offer unions power of initiative. Negotiations in Paris secured some delays—both a one-year reprieve (1980 instead of 1979) for a narrow strip mill and the agreement to keep a coking plant in Longwy, for example—but they accomplished little else in the industrial arena. Centralized collective bargaining brought the array of measures to ease workers out of the industry. These were important feats in their own right, but they did not meet the aspirations of either the CGT or CFDT. They were last-ditch defensive and temporary measures. Most important they neither protected the integrity of the worker communities nor smoothed the transition to high-paying stable jobs. Consultative bodies did not compensate for the absence of institutionalized power. The industry and the state formulated industrial policy, while the unions were relegated to executing social policies decided from above.

The absence of union responsibility weakened the authority of the federation in relation to the rank-and-file. This is evident in plant-level voting behavior that assumed a greater volatility in the 1980s. The long-term trend after the 1950s had favored the CFDT at the expense of the CGT. The two had reached rough parity in workplace elections in the Lorraine steel industry by the late 1970s, although the CGT maintained its leadership role in the Nord and the Centre. The CFDT even surged ahead at the SMN plant in Normandy and at Solmer-Fos. A decade later, however, plant-level voting patterns had become fractured. Unions gained or lost support based on their tenure on works councils during employment downsizing. In individual shops and job sites, the fluctuations resembled the fate of political parties.

Certain trends were evident, however. While the CFDT tended to run works councils and plant councils, they did so in alliance with the FO, CFTC, and CGC. Meanwhile, as the counterforce to the industry's participative management, the CGT increased its popularity. It regained its position as lead union but did not staff most works councils in the industry. The election results for union administrators in 1989 demonstrated the changes that had taken place in terms of employee representation. The CGT surged ahead to collect 47 percent of the votes.[74] Where production was concentrated—Fos and Dunkerque—the CGT did well, while the CFDT maintained an edge at the Unimétal plants in Lorraine even in 1990.[75]

Ironically, some union leaders made the jump to management. At the national level, both André Sainjon (leader of the CGT-FTM) and Georges Granger (head of the CFDT-FGMM) left office in 1988 and took positions as management consultants. At Usinor-Sacilor, the deputy director of communications at Sollac-Florange and a number of middle-level managers in Social Affairs at Sollac-Fos were former CFDT militants. These managers saw their newfound responsibilities as similar to those they held with the union—protecting workers, developing human capital, and maintaining employment.[76] These transitions also indicated a relatively porous boundary between management and labor, at least partially the result of the latter's administrative responsibilities.

Thus, the French state assumed the paternalistic role formerly held by the employers. It guaranteed employment from 1966 until 1977. When such guarantees became impossible in the face of industry losses, it assured the orderly exit of workers with the increasingly lucrative measures discussed above. Like the employers' policies of an earlier period, however, the neo-paternalistic state refused the devolution of responsibilities to the unions. Industrial policy remained the prerogative of state actors. The unions could administer but not formulate labor policy.

Conclusion

The steel case has shown that policy failures can be as grandiose as the original intentions of policy makers. Industrial policy in steel, if measured by the length of time and financial expenditures necessary to make the industry competitive, cannot be considered a success by conventional yardsticks. Part of the policy failure in steel rests with the influence held by market actors who shaped the use of state-managed resources. The steel policy-making community created a logic of its own that proved destructive to the long-term competitive capacity of the industry. Earlier state control of the industry would probably not have avoided the overinvestment and productivism that undid the industry since state intervention later perpetuated these practices until their failure was too blatant to ignore.

State-led adjustment succeeded only partially in insulating policy making from social pressures. State and industry actors failed to remove politics from investment decisions because the substance of policy directly affected large groups of workers and the economic viability of entire regions. The exclusion of unions from decision making only embittered and emboldened organized labor. Its "voice" was heard through protest.

In earlier periods, mobilization could be quelled by police power.[77] The strikes of 1947 and 1948 were marked by the intervention of both the police and the army. Although coercive power was wielded in later periods, it was used more to keep order than to break strikes. The democratic state finds overt coercion counterproductive in both electoral terms and conflict resolution. Yet because threats to public order destabilize democratic politics, they demand a response. State intervention proved necessary to ensure social peace and to facilitate industrial adjustment. The more the state became involved in industrial change, the more it needed to remedy its effects. This became even more evident as those effects blared from newspaper headlines and evening news reports. Industrial unrest consequently threatened the legitimacy of the political regime. This threat became all the more acute as electoral outcomes hinged on policy-making successes.

The interventionist state, however, cannot rely on the outcome of independent bargaining. Because it funds the social measures that make agreements possible, it seeks to contribute to the content of those agreements. Moreover, it rarely trusts management and labor to reach agreements consonant with state policy. Consequently, both management and unions look to the state for direction, encouragement, and resources. Thus, the French state cannot be a disinterested actor nourishing the organizations of labor and capital. On the contrary, it monitors every negotiating position. Collective bargaining agreements over the terms of redundancies need its approval.

Consequently, the state found itself in a double bind in steel politics. It needed to resolve conflict for both its own political legitimacy and for the sake of the industry's competitive capacity. Yet its own reflexes (not to mention those of the *patronat*) precluded union participation over the terms of industrial change. It wanted what it could not have: fruitful bargaining that would defuse a tense social environment. The resolution of this double bind consisted of buying social peace and enabling unions to bargain over those issues unrelated to product market decision making. The cost of social peace has been more than monetary. In the process of funding social measures, the state has weakened the mediating capacities of union organizations.

Although oppositional union strategies fit conveniently into this picture, they were not determined by state-industry exclusion alone. Rather, a labor movement whose reflexes were grounded in a class perspective of capitalist society found the vindication of its perspectives in its own exclusion. An increasingly entrenched (until the 1980s) Communist Party nourished oppositionalism. Still, the fractionalization of strategic orientation in the 1980s demonstrated that exclusion alone did not determine oppositionalism. A large segment of organized labor has sought negotiation despite more subtle forms of exclusion—a participation limited to social issues.

For most of the postwar period, a protest model of industrial relations accompanied state-led adjustment. Protest resulted not from frustration or alienation (although these were present) but from a strategic calculation of interest. By invoking state intervention, the institutions of labor sought to challenge the process and the substance of industrial policy. Unions lost the procedural battle and won the substantive one. Substance meant little in the face of market hemorrhage. The major goals of steelworkers have been attained: nationalization, centralized bargaining, a *statut* of public-sector status for the work force. Yet without a procedural position, French unions have been incapable of deflecting the effects of the international steel market on its own industry.

In the 1980s governments departed from the pattern that had characterized earlier forms of French industrial adjustment. They combined policy tools to restore competitive capacity. In the process, they involved unions more in the administration of labor policy. The policymaking process weakened the social bases of opposition. Instead of arresting or excluding the opponents of "modernization," the state "converted" them.

NINE

The Denationalization of Industrial Politics in France

As the adjustment process in French steel reveals, product market and labor market policies reflect political choices made within complex and constantly changing constraints. Analyzing ongoing interactions among the state, business leaders, and labor generates insights that cannot be achieved by focusing on a single player at a particular point. Detailed examination of the steel sector offers a nuanced picture of the major actors' strengths and weaknesses vis-à-vis other segments of the policy community. It also clarifies important internal debates among state planners, industrial managers, and labor leaders. A dynamic view of the adjustment process draws attention to shifting power resources, strategic learning, and preference adaptation. Of equal importance, it demonstrates that influencing one aspect of adjustment often has unintended consequences in other areas of product and labor market policy.

The great irony of the French steel story is that the state matched central control of product and labor market policy only after the national industrial model had become obsolete. This raises three broad questions. First, what does the steel case suggest about traditional patterns of adjustment, inclusion, and protest in French industrial politics? Second, why did the state use national policy tools to increase both the autonomy of firms and international integration in the 1980s, thereby diminishing central control over industrial adjustment? Third, what do past practices and current conditions suggest about how the French state, business leaders, and union leaders might adapt to industrial politics in the 1990s?

This chapter addresses these questions by placing the steel study within a wider pattern of French interest aggregation and policy making. The basis for postwar adjustment depended on the state's willingness and power to intervene in product markets. The policy U-turn in 1983 by the governing Left brought not only budgetary austerity but also new modesty about the state's role in the economy. European integration was critical to this new conception, although it was accompanied by a simultaneous decentralization of the adjustment process. The changing locus of policy formation weakened a pattern of interest representation that had accompanied the national model. Although trade unions have sought to counteract such problems, the logic of competitive unionism has exhausted organizations and facilitated a firm-based adjustment.

Adjustment, Inclusion, and Protest in Sectoral Politics

The adjustment process in steel highlights the use of political resources and the distribution of political power *over time*. The policy community that took shape to confront market failure contained a strong but vulnerable state, a trade association that tried and failed to mediate company interests, and a divided labor movement seeking to assume some responsibility. Along with the state's enlarged role came direct and strategic pressure from labor on public policy. At first, the state protected employment. Later, it helped the industry return to profitability, bypass the unions, and focus on a market-based social policy. Public policy bought competitiveness for the industry, but at a high cost for all concerned.

By the late 1980s, international competitiveness required an unambiguous response to changing world markets. The dispenser of investment funds needed to impose coherent manufacturing solutions on local managers. It needed to depoliticize the employment relationship and to alter the adversarial relations between business and labor. These exigencies were met within the context of political choices and market constraints. The French state succeeded through support from Brussels, an ideological transformation within the Socialist majority, the development of new policy tools, and the collapse of union oppositionalism.

Industry representation changed considerably in the forty years after the inauguration of the ECSC. In the early postwar period, the CSSF had a monopoly on representation, but it was not an independent actor. The steel firms pursued individual investment and production policies. Voting within the trade association was based on sales, and the Lorraine companies dominated decision making. The CSSF saw itself as the industry mouthpiece, reconciling divergent interests, but *not* enforcing decisions made in negotiations. The steel companies tried to maintain a way of life, but that quest proved futile given the increasing openness of French markets. The organization therefore could not take decisive autonomous action.

The demands of international competitiveness and technological change forced the industry to choose between market decline and state aid. Industrialists gradually relinquished decisions over production, investments, and eventually labor policy to the CSSF which, in turn, secured resources from the state. Market breakdown altered sectoral pluralism such that the trade association gained considerably greater authority. The governance of steel politics devolved to the CSSF partly because it could extract resources from the state, partly because the state wanted a single interlocutor, and partly because a new generation of leaders repackaged the trade association's internal resources.

Market hemorrhage ended the political legitimacy of the close business-state relationship. With nationalization, one version of the steel policy community was destroyed. Companies came under the authority of the state, and the trade association was divested of power. Public ownership, however, made the tensions within the policy community more transparent. Many regional rivalries among the firms became apparent as all competed for resources in a shrinking market. The 1984 debate between Lorraine and the Nord over the siting of a sectional mill—either by expanding Gandrange or building a minimill in Valenciennes—separated ministers in the Socialist government and divided the labor movement.[1] The connection between investment and employment was clear.

The position of labor complicated the adjustment dynamic. Initially on the outside of the steel policy community, the unions had relied on protest to influence policy. By the mid-1960s, that protest secured limited victories for the labor movement: no forced layoffs took place after 1966 in companies constantly facing bankruptcy. Instead, the instability of industrial relations seemed even at times to favor union interests in bargaining: between 1966 and 1977 French steel workers saw improvements in both wages and employment security. Labor learned to use protest as a bargaining lever.

As employment problems became acute, however, the steel unions were increasingly invited to participate in administering labor policy. This participation involved programmatic transformations for the unions, which in turn created internal conflicts over the appropriate role for labor in the industry. Such labor involvement could take place only because the balance of power shifted away from the CGT and toward the CFDT. As the unions administered both passive and active labor market programs, the companies tried to alter existing employment relations at the workplace—but without union input.

Industrial adjustment was an intensely political process in the French steel industry. Both business and labor sought to influence state policy toward steel. The trade association and the labor unions tried to use the state to protect markets and extract resources. Unions differed over the role for labor and industrial policy, while the state attempted to change the behavior of business and labor. Each could claim only partial success.

Consequently, the so-called strong French state had difficulty implementing steel policy. In the 1950s, it encouraged investment but could not compel the uniform dispersion of new technologies. In the 1960s, it insisted on mergers but could not prompt firms to shed older facilities. In the 1970s, it forced a financial solvency plan but could not create the necessary manufacturing efficiencies. In labor markets, the state modernizers acquiesced reluctantly to firm-level autonomy, for they feared jeopardizing their industrial policy. That acquiescence complicated the state's relationship with organized labor. The state sought to encourage bargaining, but the unions and companies were reluctant and suspicious partners. By the 1970s, the state's only labor policy tools consisted of expensive programs to pay for exit from the industry. Government policy makers finally succeeded in articulating labor policy with industrial policy in the 1980s.

The history of French steel adjustment shows the extent to which economic actors change policy preferences over time. Management, which defended product market autonomy in the 1950s, embraced a joint effort with the state in the 1960s. The managerial opponents of European cooperation in the 1950s became its ardent defender in the 1980s. Likewise, within the labor movement, the same unions that sought state control in the 1950s criticized the use to which that control was put in the 1980s. Of course, new individuals had assumed leadership roles over the forty-year interlude. Likewise, those policy perspectives responded to very different economic environments. Nonetheless, organizational interests were transformed.

This instability of interest articulation and representation suggests that political actors can use market resources to reinforce their own political assets. Within product markets, the CSSF controlled the flow of market information to the state to protect the industry from potential intrusion. In conjunction with the poor economic performance by the companies, the trade association became the kingpin of the steel policy community. Yet the creation of Usinor and Sacilor—embodying the success of the national champion policy—encouraged the state eventually to cut the CSSF out of the policy process.

Within labor markets, the unions pursued an opposite course: they used political leverage to multiply market resources. The unions used the threat of social unrest as a lever to increase their involvement in the policy process. Virtual security of employment boosted wage gains. Moreover, aid to steel companies politicized the employment relation. Politicization served as a resource to compensate for the organizational weaknesses of the unions.

For the state, market and political resources had a reciprocal influence. Inspired by a critique of economic performance, modernizers sought to organize the market. Market failure in steel, however, pressured industrial policy makers to become even more involved in the adjustment process. Yet policy leaders were also impelled to market action by the extent to which social interests could marshal political resources by mobilizing public opinion.

This intersection of markets and politics in the steel adjustment process accounts for the incomplete and temporal nature of outcomes for business, labor, and the state. This process had no "winners." The industry became internationally competitive but at the cost of managerial autonomy. The unions gained inclusion in regulating the quantity of labor but were marginalized from the more dynamic process of encouraging labor loyalty to production. The state usurped control from business but at enormous costs in general revenue and in the legitimacy of the policy process.

Thus, steel adjustment clarifies answers to questions about industrial governance raised in chapter 1. The claim that some actors are included while others excluded from policy making needs dramatic revision to take into account both symbolic inclusion and adaptation to formal exclusion. Moreover, interest representation depends on political factors—mobilization of ideas, leadership skills, agenda setting, and party competition—which interact *dynamically* with market forces. Finally, protest can consistently affect policy making by altering political agendas.

The National Focus of French Industrial Politics

The steel case differs from the broader pattern of adjustment in the postwar French political economy in that serious structural problems were evident by the early 1960s. Yet both the steel sector and the larger economy adjusted within clearly discernible national borders.

Throughout the postwar period, the nation-state was the locus for economic decision making. The international dimension of policy consisted more of securing a national basis for production than of internationalizing economic activity. Efforts to integrate the European economies via the European Recovery Program, the ECSC, and the European Economic Community were undertaken largely for geopolitical reasons. Struggles over resources were domestic affairs.

Such a domestic focus was not extraordinary in the early postwar period.[2] Various indicators—ranging from levels of foreign direct investment to trade as a percentage of gross national product—showed that economic policy making had a national basis in all medium-sized to large industrial countries.[3] A prosperous consumer market would drive economic growth through a multiplier effect. Development of the strong state led the French political community to believe in its ability to constrain market forces politically. This voluntarism, together with the policy tools of the positive state, encouraged public authorities to pursue actively interventionist industrial policies.

France's economic growth depended on both fortuitous convergences and the interventionist state. Few would have predicted that a backward-looking country of baguette-carrying peasants, complacent financiers, and comfortable political notables would have mobilized capital for the "thirty glorious years" of economic expansion. Yet after the war the lock of the rural sector on the French economy was broken. More important, industrial growth took a startling turn. Productive investment tripled in constant prices between 1949 and 1966.[4] Between 1958 and 1980, France and Japan enjoyed the highest growth rates in the OECD area. France used the markets made newly available by the

Common Market to produce a version of export-led growth. The steel industry lagged behind other sectors, for while investment was high, production and productivity growth rates were slow—due to a failure to retire surplus capacity.

Economic management depended on the financial tools wielded by state planners from 1945 to 1980.[5] Productive investment, in turn, spurred steady growth. While France had previously experienced high rates of net capital formation, they were usually followed by cyclical downturns that destroyed the gains in productivity.[6] From the 1950s to the 1970s, there was continuous and uninterrupted capital formation. Partially as a result, debt-to-equity ratios rose during the postwar period, both creating dependencies on state finance and giving the state levers of influence.

State intervention in financing investments facilitated the transfer of resources from backward sectors to advanced ones. Less and less sheltered markets forced French producers to alter their old practices and adopt new ones. Policy makers used the opening of European markets to force business to achieve German standards of productivity. Thus, political will assumed greater importance than autonomous market forces in forcing market actors to invest and produce. Ironically, basepoint pricing in the ECSC delayed that demonstration effect in steel.

French capital became more risk-taking and export-oriented, especially in key sectors of mass production. Like their counterpart in steel, industry associations opposed the internationalization of the French economy during the 1950s.[7] Like their steel industry peers, the new French managers were forced to adapt to market competition. Economies of scale lowered unit labor costs. The rural sector was transformed from self-sufficiency to an important export sector. The peasants became capitalists.[8] As a result, economic activity had changed radically since prewar times, thereby transforming cultural attitudes toward growth and accumulation.

State enterprise also acted as a locomotive for the economy. In the post-1945 period, the public sector served as a model for allocating resources. State control of the banking sector, for instance, facilitated the distribution of credit. The state productive sector—companies such as Renault and the coal mines—also set the pattern for collective bargaining.

The French economy was distinguished by a dual business structure. As elsewhere in southern Europe, governments used the political support of the small business sector for electoral purposes while distributing economic gains to advanced sectors.[9] The family firm decreased slowly in importance, and this was true both for large and small business.[10] Still, by the 1980s, local labor markets were dominated by one or two firms. This is not to denigrate the importance of porcelain in Limoges, the scientific community in Isle-de-France, perfume on the Côte d'Azur, or clock making in Besançon. Rather, it suggests that in France, more than in Italy, the small firm has lived in constant tension with the large enterprise.[11]

Therefore, cleavages developed between state-owned and private firms. In the public sector, the firm was the model of American-style rationalities. The private sector combined those rationalities with French traditionalism. This was part of a larger bargain to which organized labor could do little but submit—high growth but social backwardness. French employers were culturally predisposed to exclude labor. Until 1968, public authorities did not attempt to alter this predisposition.

Wages rose during this period of economic growth, but labor did not participate in redistribution.[12] Unions could sit on the sectoral commissions of the Plan, but investment decisions were made elsewhere. The marginalization of labor strained the legitimacy of the policy process, but it obviated the necessity to iron out bargains between unions and political parties to regulate the economy.[13] Mass production and a self-sustaining circuit of growth developed in the "golden age" without the active involvement of labor representatives.[14]

A set of mutual dependencies developed within French adjustment. Unions needed state intervention to compensate for their inability to negotiate with employers. State planners needed the policy levers of the interventionist state to maintain levels of consumption, stabilize labor markets, and maintain social peace. Employers came to rely on (or at least expect) labor market intervention, given their unwillingness to tolerate a counterauthority in the labor market and their dependence on state intervention in product markets.

Economic growth depended fundamentally on a national mode of adjustment. After 1950, state-sponsored investment, technology transfers, and organizational methods encouraged the growth of mass production industries within French borders, although they depended increasingly on foreign markets for economies of scale. The policy of national champions concentrated industrial assets in a small number of large companies. French economic growth benefited from expanding European and U.S. markets and a franc pegged to the dollar.

International markets changed for France by the 1970s.[15] The rapidity of growth under the stewardship of de Gaulle and Pompidou led to a dual economy of mass production industries and small businesses, capably coordinated in political terms but increasingly vulnerable to market turbulence. The engine had been the national champions forged in the fifteen years after 1958. Such champions depended on increasing economies of scale and expanding markets for their goods. This model encountered difficulties in the 1970s. Export dependency was vulnerable to the low-wage mass production of the newly industrialized countries. The stability of mass production was upset by turbulence in the international trade and exchange rate regimes. Innovation, productivity, and job creation began taking place disproportionately in small firms. The strategy of producing for national consumers was less viable with more open borders in Europe and the initiatives on the Single European Market of 1992.

The *national* model of industrial adjustment placed ever increasing demands on central policy makers, who often lacked the political autonomy or market control to deliver on their promises. In both the steel industry and the wider economy, such political demands encouraged new thinking among policy leaders.

Market-Driven Adjustment

By the time the French state acquired the product and labor market tools that modernizers had envisioned, central control was obsolete. Already in the 1970s, the politicization of industrial and labor policies came under attack from different directions. For some observers, policy making was becoming increasingly difficult, given the demands for participation.[16] For others, this translated into critiques of the weightiness of the French state and dissent over the adjustment process from within the Left itself.[17]

By the mid-1980s, moderate Left and Right political leaders concluded that detailed intervention in targeted industries was both expensive and counterproductive. They sought competitiveness for firms in international markets to stimulate economic growth and political stability. The logic was that firm-level profitability would mean distributional gains for employees.

Economic policy after 1974 alternated between stimulating demand

and generating investment. Both failed on their own terms. The attempt by the Chirac government to infuse purchasing power in the economy between 1974 and 1976 stimulated inflation and aggravated the external balance; it did not rekindle the growth cycle. The attempts by the Barre government to carve out market niches for the export sector and to use a strong currency to sort out the fortunes of French firms left French industry relatively weak in uncertain European and international markets.[18]

The attempts by Giscard and Barre in the late 1970s to reform state-society relations stumbled on the resistance both of a business community accustomed to its privileges and a labor movement unwilling to relinquish its vision of a better society for market institutions.[19] Both business and labor were polarized in the wake of the two oil crises. Given such market failure, the state became an even larger site for distributional politics. Industrial policy in the 1970s consisted largely of extending subsidies to large and small firms that suddenly found themselves in competitive difficulties.[20] Labor market policy had to cope with the pressures for redundancies. The state's arsenal of tools brought increasingly expensive programs to pay workers not to work.

With the election of François Mitterrand as president and the overwhelming Left majority after the subsequent legislative elections, the Left put into practice the policies developed while in opposition: the nationalization program promised more political control of the economy, labor law reform (the Auroux Laws) would democratize industrial relations by offering opportunities for "industrial citizenship," public policy would be decentralized. Against the current of the other advanced countries, the Left government stimulated the economy by raising incomes and consumption.

The Left in 1981 sought to revive national champions (through vertically integrated *filières*) and to inject purchasing power into the economy. The history of the reform experiment has been copiously chronicled, so we need only suggest reasons for the U-turn.[21] Demand stimulation fueled the consumption of imports over domestically produced goods, thereby failing to produce enough demand to spur industrial recovery and putting pressure on the external trade balance.[22] The frailty of the French economy was exposed in the red ink suffered by both the new public sector and private companies. The underlying weakness of French industry in general contributed to high import elasticity. In mid-1982, the government froze wages and prices, later deindexing public sector wages. Devaluation sought to slow the run on the franc in October 1981 and June 1982. Another round of measures in early 1983 transformed a "pause" into austerity: wage-price guidelines reduced purchasing power, budgetary retrenchments stemmed increases in transfer payments, and another devaluation sought to restrict consumption of foreign goods. The Socialists made a dramatic reversal less than two years into power.[23]

Since 1983, French economic policy has been characterized by careful economic management and policy orthodoxy. Despite three changes in legislative majority and seven different governments, the goals of economic policy making have remained consistent after the U-turn in 1983—a strong franc, stable prices, firm-level profitability, and renewed investment.

Orthodoxy departed from the economic policy characteristic of the postwar model. During the period of growth, demand stimulation—through increases in consumption or investment—was constantly balanced by periodic devaluations. After the mid-1980s, in contrast, economic "rigor" dampened collective consumption. National budgetary stringency kept consumption below increases in national product. France even began reducing the national debt, facilitated partially by the proceeds from privatization in both 1986–1987 and after 1993. When collective expenses became unavoidable, as in the case of the social security system, households increased their contributions.

This policy shift increased business confidence and produced an environment more favorable to investment. Yet the state has also done its part to encourage domestic investment. Decreases in corporate profit and value-added taxes have lowered costs at each level of production. Employer payroll contributions were stabilized after fifteen years of continual increase.

Another inducement to business confidence was the diminished threat of the intrusive state. In the early 1980s, municipal governments began contracting out some services. The Chirac government privatized some firms, while the 1987 crash and the 1988 election defeat put a damper on further action. After the five years of "*ni-ni*" (neither more nor less nationalization promised by candidate Mitterrand in 1988), the Right government elected in March 1993 promised a wholesale resumption of the privatization program.

The stable currency facilitated the coordination of European economic policy, positioning the French economy, so policy makers hoped, to compete on an even basis for the single unified market of 1993. The strong franc enabled French companies to diversify on a European level. France no longer risked capital flight to countries with stronger currencies—Germany, for example.

The strong franc also altered French labor markets. A high value for the franc meant that domestic interest rates needed to remain high to attract funds to France. This interest rate policy dampened economic growth and kept unemployment high. Employment policy had a disciplinary effect on wages, as real wage growth lagged behind increases in national product. The decade of bipartisan orthodoxy that began with deindexing wages in 1983 continued into the early 1990s with various programs, affecting millions of workers, that either encouraged employers to compete on the basis of wages or compensated an increasingly dualist human resource strategy.[24] Even the CNPF admitted that increases in French unit labor costs had dropped to the second lowest in the European Community.[25] A government report from 1990 highlighted growing disparities of wealth and income.[26]

In order to tackle the dilemma of abundant workers with insufficient skills, French governments embarked on an ambitious path to reorganize employment and training programs. The new consensus on adjustment translated into a jaundiced view of passive labor market instruments that were perceived as expensive and ineffective.[27] Creating wealth and encouraging technological innovation in the firm demanded more proactive policies. Workers needed training for the new work environment.[28] Employment services had to be reorganized to meet those tasks.[29] After 1987, the Ministry of Labor developed policies to offer employers incentives to hire additional workers and to promote workplace training.

This adoption of active labor market policy, however, came with a twist. The training opportunities took place at the plant level—at least partially because of the insufficiencies of the state vocational training system. Employers received relief from social security contribution as an inducement to hire workers. In raw numbers they were successful. Still, many of the jobs were short-term—becoming dead-end in terms of the individual's career—and workers tended to move from one program to another. Other programs offered young workers part-time employment at sub-minimum wages. In general, these employment programs received the label of training, although little state oversight was imposed. The training component has been minimal and job-specific as a result. Employment subsidies have effectively induced French employers to compete on the basis of low labor costs, although they have the *political* benefit of lowering aggregate unemployment figures.

Governments after the mid-1980s also introduced flexibility into the hiring, firing, and the use of labor. Chirac abolished the procedures whereby employers needed administrative authorization for layoffs, and his Socialist successors refused to restore it. Subsequent legislation on layoffs only provided for more consultations with works councils. This has effectively reinvigorated firm-based human resource management.[30] Already in 1986 the Loi Delebarre permitted flextime by enabling employers to use annual (rather than weekly or monthly) work hours to juggle work schedules. Governments after 1984 have been more lenient toward various forms of precarious work.[31]

In the midst of financial incentives to restructure and cap wages, French companies have also been in the forefront of attempts to instill new forms of cooperation in the workplace. The five Auroux Laws of 1982–1983 gave workers a set of rights within the workplace. Employees received the "right to expression." Works councils had to be consulted in the event of major changes in personnel. (And they received independent powers to assess management plans). Firms had to bargain with unions on an annual basis. Health and safety committees received expanded powers to stop production. Workers for companies in the public sector could elect one-third of the board members.

This attempt to institutionalize workplace citizenship decentralized labor-management relations. Wage negotiations shifted to the plant level. A number of companies also sought new ways to involve employees in the fortunes of the firm. Quality circles were the rage of the early 1980s, and the law requiring employee direct expression circumscribed the counterauthority of unions at the workplace. The concern with training encouraged management to cede some human resource management to works councils and independent subcontractors hired by the latter. Therefore, the centralization of collective bargaining—in steel and elsewhere—that had occurred in the 1970s withered a decade later.

Market-driven adjustment did not entail less state intervention in industrial or labor market policy.[32] It showed instead the extent to which the state could mobilize for new ends. The intent of budgetary austerity, wage restraint, labor market flexibility, and stimulation of business confidence was to attract high-quality investment to France. This was not expressly a low-wage strategy, although some companies interpreted the new labor market in that way. Rather, economic orthodoxy sought to attract investment for a European strategy of production, sourcing, and consumption.

The European Solution

As a medium-sized country, France has had less maneuverability in confronting international instability. Political leaders concluded by the mid-1980s that economic policy needed to be framed within a broader European framework.

Building a more coherent European Community took place in four stages for French policy makers. The first stage in 1983, the choice of exchange rate policy (whether and how much to devalue), was a struggle over domestic economic policy and relations with Germany.[33] Several members of Pierre Mauroy's government argued that France should leave the European Monetary System (EMS) to give the economy more room for maneuver. Mitterrand apparently seriously considered breaking with the EMS and even decided to take the leap.[34] Yet he reversed his course when econometric projections promised a massive loss of foreign reserves and an improved trade balance only in the medium term.

The decision to remain in the EMS set the terms for economic policy throughout the decade.[35] It also seems to have been critical in marginalizing those on the Left who supported autarkic policies (the so-called Albanian solution) and in reinforcing Mitterrand's pro-Europeanism. The prominence given to the discipline of exchange markets decreased the Left's ability to carry through its electoral promises of redistribution. It also put French policy under the tutelage of the Germans: Finance Minister Stoltenberg demanded the deflationary measures in exchange for the devaluation. Finally, it paved the way for the rush toward the Single European Market by guaranteeing stability on the French side of the Bonn-Paris axis.

The second stage in the European project was the Davignon Plan to restructure the European steel industry. The choice to submit to the authority of the European Commission in 1984 marked an important step for French industrial policy. As suggested in chapter 7, Community policy probably favored the weak French producers even during the crisis programs. Still, the *dirigiste* steel policy moved steel policy making from Paris to Brussels.

This displacement of product market initiative was not simply an attack on national sovereignty, however. It signaled an attempt to use market mechanisms to sort out a large and politically sensitive sector. The Socialist government proved willing to reverse earlier policy and to withstand public protests from organized labor. The performance of Usinor-Sacilor provided a concrete and positive example of European self-help.

The third stage was the push behind the Single European Market. The Single European Act (SEA) of 1987 owed much to the activism of Mitterrand.[36] As president of the European Council between January and June 1984, Mitterrand spearheaded efforts to forge a compromise with Britain over its contributions to the European budget and with Germany over expanded membership to Spain. Jacques Delors, his appointment as president of the Commission, worked quickly to forward the Dooge Committee report to national governments recommending liberalization of the internal market. Lord Cockfield's White Paper concretized that recommendation into 287 proposed measures to initialize liberalization. A Franco-German alliance pushed Britain into accepting a procedural reform in the Council of Ministers that would substitute qualified majority voting for unanimity over those directives. Between late 1983 and the 1987 adoption of the SEA, Mitterrand forged a new European compromise.

The Maastricht Treaty for Economic and Monetary Union (EMU) signed in 1991 constituted the fourth and final pillar of French European policy. Like the SEA, it was vigorously pushed by the French. It appears to have been an attempt to tie Germany to Europe after unification and to avoid continued dominance by the Bundesbank. The proposed European Central Bank System implied that the Bank of France gain independence from the Ministry of Finance. This had the support of both the Bérégovoy and Edouard Balladur governments, although it created tensions within both coalitions. The limits on inflation rates, long-term interest rates, and budgetary deficits that constituted the convergence conditions for implementing the EMU gave French policy makers even less fiscal and budgetary discretion than they enjoyed after the U-turn of 1983. The EMU brought to fruition a decade of denying the efficacy of Keynesianism and state voluntarism, thereby diminishing control of the economy by party politics.

The French policy on Europe was a startling turnabout for a state-centered polity. The SEM limited the state's temptation to intervene and reduced the autonomy of domestic policy making. It extended the exchange rate conformity of 1983 into other policy domains—standard setting, regulations, and state intervention. It expanded the market contours facing European economic actors. Consequently, French firms have sought a European *space* for investment, production, and marketing. While it remains uncertain how much decision making was sacrificed in Paris, it is clear that product markets have become more European.

The activism in the choice of Europe seems to have come from a recognition that France could not compete on a national footing.[37] The U.S. recovery had started later than the French hoped, but job creation in the United States after 1983 stunned French observers who were bemoaning net losses in France. The continued success of the Japanese sent shivers up the spines of policy makers who envisioned Europe as the next landing site after the American market for Japanese firms.

French firms have taken advantage of the new terms of competitiveness in the French and European markets to expand outward into Europe and beyond. In the late 1980s, profits were high in both the public and private sectors. French investment in the United States tripled between 1987 and 1990. French capital bought up the water authority in Britain and the Saar steel industry. BSN, Bull, Michelin, Péchiney, Renault, Thomson, and others entered strategic alliances at the European level. The internationalization of French business affected both the public and private sectors.

Such foreign investment has given French companies considerable leeway in the European market. It allowed French companies to capture market share, tap into local skills, and spread risks across the continent. Its impact on organized labor was less positive. Most of these companies had trimmed their work forces by the mid-1980s, although some companies could both buy up foreign assets and lay off workers in France.

The attempts to recast Europe have made French industrialists into more autonomous players in the adjustment process. While not released from state supervision, large French companies have a greater capacity to look elsewhere for capital. Cross-border mergers and strategic alliances have given firms the freedom to forge their own investment choices. Smaller companies benefit from the liberalization of finance to the extent that European competition in banking and securities has made national policies difficult to enforce.

Labor and Firm-Level Politics

French trade unions have not been able to respond effectively to the new pattern of adjustment for three reasons. First, quality production altered the terms of employment, and the unions have not been able to mobilize new categories of workers. Second, the state intervened in labor markets to ease workers out, further demobilizing the work force, and orthodox economic policy precluded broad labor market stimuli. Third, competitive unionism meant that efforts to confront the first two problems would not be coordinated.

It is easy to be pessimistic about French unions both inside and outside the steel industry. They seem more disoriented and strategically uncertain than at any other stage in their history. The search for flexibility has reduced the aggregative capacities of unions. Workers tend to view union organizations in very instrumental terms as evidenced by the volatility in election returns for workplace representatives. Membership, already lean in 1980, was at least half that number ten years later. Public statements suggest a density level of 9 percent of the work force. Discussions with pessimistic union leaders elicit even lower figures. In the early 1990s, France had the lowest proportion of organized workers among the industrialized countries.[38] Membership is distributed among the "big five" confederations and other smaller unions.[39] In spite of the different meaning of membership in France, such figures convey real changes in the representation of workers' interests.

The embrace of market norms by the French state widened the preexisting divisions within the labor movement. The fallout over propositional unionism and *recentrage* was part of a larger reconfiguration of the French union movement in the late 1970s and early 1980s. For the CGT, the CFDT had succumbed to reformism, while the latter claimed the former had simply reverted to its earlier position on the transmission belt to the PCF. The rift grew wider in the next fifteen years. Even symbolic displays like May Day parades occasioned such vicious displays of animosity that each confederation marched separately. Works council elections provided the opportunity for alliance building to exclude other unions. Employers' reprisals against organizers from one union were typically greeted with silence from competing unions. While some cooperation took place at the local level, the confederations also attempted to reimpose discipline. Both the CGT and CFDT used expulsion and decertification of locals to gain control over local actions.

Given the aggravation of divisiveness within the labor movement, the three major confederations began forwarding different policy responses for the new competitive environment. The CGT advanced wage demands whose realization was usually unlikely. As in the 1950s, it used "hard" positions to advance a criticism of society, with the major difference that mobilization around alternative projects became much more difficult. It continued to advocate a Keynesian response to unemployment—wage hikes to stimulate consumption.

Meanwhile the CFDT showed a willingness to become "responsible" and participated in what best can be called concession bargaining. It was in the forefront of firm- and national-level bargaining. Its former secretary general termed the strike an "archaic" tool of the union movement. It formulated a confusing array of initiatives that ranged from new growth models to increased training to an efficient public service.

In the initial stages of this feud, the FO was the prime beneficiary. It portrayed itself as above partisan conflict (while lambasting the Socialists for inviting the Communists into government) and focused on doing what it always did—bargain for more. Its electoral returns increased somewhat in the early 1980s, and while membership in the other two unions plummeted, FO rolls remained stable. Still, the employment hemorrhage in the mid-1980s affected all unions, including the FO.

The FO later found itself riven by its own internal cleavages and bounced between its previous role as key signatory to a more contestational unionism. It supported the CGT's refusal to sign framework accords in the civil service and a national accord on precarious work, both signed by the CFDT. Yet it tended to sign firm-level accords, frequently in complicity with the CFDT. Nonetheless, it criticized the centrist economic management after 1983 and the eroding purchasing power of its members. If the CGT reinforced its class position and the CFDT emerged within the centrist coalition, then the FO articulated a new labor market populism.

The CGT had long since lost its hegemony within the labor movement. By the 1980s, it was struggling to maintain its position. Until the leadership change in the FO in 1989, the CGT frequently found itself outmaneuvered by informal alliances between the CFDT and the FO. This was true in national-level negotiations over workplace flexibility (ultimately undermined in 1984 when the membership of the CFDT disavowed its negotiators). The CGT negotiated actively but had less propensity to sign agreements than its rivals.

Certain observers have found in these anti-CGT coalitions the basis for a new push toward labor unity—minus the CGT. The late 1980s witnessed a bevy of proposals for reconfiguring the labor movement. In the coal mines, the CFDT, FO, CFTC, and CGC developed coordinating mechanisms to negotiate salary accords. In Lorraine, these four organizations coordinated positions in the administrative mechanisms to ease out redundant steel workers. Similar coalitions existed in the banking sector, the public sector, and among teachers.

Yet the dreams of a new social democratic (noncommunist) labor movement were idle reverie, given the enduring cleavages among the unions. No sooner had André Bergeron stepped down from the leadership of the FO than his successor Marc Blondel proposed closer cooperation with the CGT, especially over wage issues. The FO, staunch in its secularism, remained suspicious of the "social Catholics" of the CFDT. Most leaders staunchly defended the concept of union pluralism either on ideological grounds or as a way to protect what remained of their organizational coherence.

Disunity in the labor movement has created a broad disaffection for individual unions. The wave of redundancies in the mid-1980s hit unionized companies the hardest. And while turnover among union members has always been high, especially given the necessity to collect dues on a monthly basis, new members have not replaced the old. The average age of union members increased substantially over the course of the 1980s. As of 1989, the largest single vote-getter in professional elections was the category of nonunionists.[40]

Changes in the workplace weakened the ability of unions to bring together different interests in the labor market. The decentralization of bargaining from sector to firm has brought wages back into competition—both by plant and by sector. Employers' attempts to encourage workers to identify with the firm confused workers' loyalties.[41] Increased recourse by employers to less-than-permanent employees (encouraged by the programs of the Ministry of Labor) also undercut the unions' ability to maintain their ranks. A young school leaver can easily circulate for years in several "training" programs without a stable labor contract or gaining marketable skills. Thus, a polarized labor market developed in which a core of well-trained workers had ties to the well-being of the firm, while a pool of temporary workers moved in and out of the workplace. Unions have historically had difficulty attracting temporary employees.

Militancy has been difficult to generate, as indicated by annually declining strike rates: by 1990 France was among those countries with the *lowest* rates (See table 9.1). In an earlier period, French unions compensated for their organizational inadequacy through local mobilization that could, in turn, trigger a response from the state. Working-class mobilization had explicitly political intentions. After 1983, it became increasingly difficult to create a large enough disturbance to compel state intervention. French strike levels for the 1980s were one-third those of the 1970s. By the early 1990s, strike rates had dropped below those in Germany.

While French labor markets have been quiet since 1981, some categories of workers—Air France employees, railway and subway workers, air controllers, postal employees, nurses, and civil servants (tax collectors!)—resisted the downward wage and employment pressures. The most effective actions have been led by *coordinations*, extraunion and frequently antiunion assemblies of workers, which have secured isolated gains. The unions have taken contradictory positions on these activities, while voicing support for wage claims. The CFDT expelled several postal locals for cooperating in one *coordination*. The CGT has sought with mixed success to control the job actions.

The *coordinations* have tended to take place in the public sector. The private sector has been much quieter. The pressure of public sector incomes policy made wage growth slower than in the private sector—the inverse of tendencies in the early 1980s. Mobilization is also easier in the public sector because the link between social costs and political solutions is more obvious.

Just when the union movement is weakest organizationally, it finds itself well entrenched in both new and old workplace institutions. The strengthening of works councils in the early 1980s, development of tri-

Table 9.1 Strike Activity in Selected Countries, 1971–1990
(annual average)

Working Days Lost in Work Stoppages per 1,000 in Total Labor Force	
Italy	1,624
Greece	454
Spain	340
United Kingdom	226
Portugal	92
United States	86
France	55
Germany	23

Source: ILO, *Yearbook of Labour Statistics*, various years.
Note: Data for Spain excludes Catalonia (1981–1986) and the Basque country (1986–1990).

partite employment commissions at the local level, and participation in the oversight of training and employment initiatives within the Ministry of Labor have given unions many more formal access points to state managers and the business community.[42] The difficulty lies in translating administrative capacity into political or marketplace power. The personalistic and ideological cleavages of the labor movement prevent such administrative tasks from becoming a mechanism to increase labor's counterauthority.

Moreover, the new rights that unions received through the Auroux Laws were not easily translatable into political resources on the shop floor. Consultation rarely results in changes in management plans. New responsibilities have also stretched already thin (and competing) union resources. Plant delegates and works council members complain of the avalanche of paperwork they must examine. Even when local union leaders can marshal technical expertise to challenge managerial authority, they frequently cannot obtain the confidence of the rank and file, who remain wedded to bread-and-butter issues.[43]

At the very time when employers sought plant-level negotiations—and until the 1980s they rejected this approach because of the potential for rapid politicization—unions were unable to negotiate effectively. They have had difficulty even processing the information to which they now are legally entitled. Increasingly, French unions are staffed only by militants, not members.

Thus, by the 1980s, the labor market had become more fragmented and less organized.[44] The labor movement tried various means to reconcile these tendencies. Some focused on increasing the national power of unions, while others tried to increase international cooperation among labor organizations. Attempts to bolster membership through such experimentation, however, could not substitute for influence, and here the record remained mixed. Bargaining flourished at the local level, yet languished at both the sector and national levels. National agreements decreased after 1981 and those negotiated in the early 1990s were intended as framework accords to be finished at the sector level. Those at the sector level sought to give firms maximum flexibility.

French labor markets were bifurcating in the early 1990s. Market adjustment, quality production, and the new employer ethos had reinvented the French workplace. With mobilization stifled and unity splintered, the labor movement had trouble presenting itself as a counterforce in society.

Europe and French Trade Unionism

Internationalization contributed to the increasing "marketness" of the French economy.[45] While this may have increased the efficiency of resource allocation, it also struck hardest against traditional forms of labor representation—national labor unions, whose major goal has been the removal of wages and working conditions from competition. A logical "societal" response would be to construct European labor institutions to compensate for national inadequacies.

Yet the national pattern of interorganizational conflict has been overlaid on an emergent European system of industrial relations. Europewide union mobilization was impeded by the continued absence of the CGT from the European Trade Union Confederation (ETUC). Its exclusion has resulted from the vetoes of the CFDT and FO, who claimed that membership in the World Federation of Trade Unions (WFTU) was incompatible with the free collective bargaining held up by the ETUC. This was an article of faith for the FO under the leadership of Bergeron. Such positioning came later, however, for the CFDT. Until 1979, it supported CGT membership. Only the feud between the CFDT and the CGT after 1978–1979 could explain the changed stance. With the breakdown of Leninist regimes, the WFTU disintegrated, and the original justification for exclusion lost its force. Although union leaders indicated the possibility of CGT membership, nothing was decided as of spring 1995.

The CGT long opposed European initiatives—from the ECSC to the Treaty of Rome to the Single European Market. It has argued that European integration has favored capital and it was "their Europe."[46] Rhetoric and the sloganeering, however, did not compensate for a true European strategy. The relatively frequent demand that French multinationals disinvest from foreign operations in order to produce in France did not enhance international union cooperation. The CGT's position on Maastricht was slightly more nuanced: when Louis Viannet assumed the leadership in January 1992 he called the accord a "challenge to workers and unions in Europe" and reasserted the union's desire to participate in Europe, but he did not condemn the treaty.[47]

The confederation expressed frustration about the limits on national policy making generated by France's European policy. It argued that constraints on price levels, interest rates, and budget deficits eliminate the possibility of countercyclical (and progressive) politics. It suggested that the economic logic of Maastricht could pressure the French to privatize the public sector and increase the likelihood of "social dumping." This was particularly compelling for the CGT as the Socialists restructured the public sector, making it more competitive and less the bastion of secure employment.

The eventual integration of the CGT into the ETUC should not induce significant policy change. Economic policy positions and its vision of collective bargaining continue to separate it from the CFDT. Ironically, the CGT has found itself in broad agreement with member unions such as the German IG Metall and the Spanish UGT, which tried to forward the Social Charter before the completion of the internal market.[48]

The CFDT enjoyed a privileged relationship with the Commission's President Delors, and its Europeanism made it a key actor in the ETUC. Its ideas filtered into the development of the Social Charter. It saw in Europe possibilities to engage in constructive, participative bargaining that had always been difficult in France. It sought to extend to the European level the new rights obtained in France in the 1980s, along with emulation of German workplace institutions.

The FO became an enigma to those studying European industrial relations. It remained staunch in its self-proclaimed Europeanism. Yet it wanted little to do with the initiatives on workplace participation, preferring to leave management to managers. As a way of reconciling very diverse tendencies within the confederation, it remained ensconced in its devotion to wage bargaining. Thus, it was a very reluctant member of the ETUC.[49]

As of the early 1990s, the ETUC's influence rests at the legislative arena. The international secretariat is small and largely subsidized by the European Commission. The industry committees have yet to negotiate a significant Europewide sectoral agreement. When they eventually complete such an agreement, the Union des Confédérations de l'Industrie et des Employeurs de l'Europe (UNICE, the European private sector employers' confederation) seems intent on maintaining national differences.

Thus, interorganizational distrust has extended to Europe, rendering French unionism ineffective as part of a continentwide movement. For instance, an agreement at the Airbus consortium in 1992 by which the staff council was constructed to represent both French workers (from

Aerospatiale) and workers from Deutsche Airbus, British Aerospace, and the Spanish firm Casa was strongly contested by the CFDT and IG Metall. The FO, CFTC, and the CGC signed the agreement, however. The CFDT and IG Metall believed that the staff council fell short of the European works council envisioned in impending legislation at the Commission. Moreover, rumors circulated that management cut a deal with these unions to marginalize the CFDT and, more important, the IG Metall. European counterparts are uncertain of France's contribution to a European collective bargaining system.[50]

Ironically, French multinational corporations were leaders in extending union rights.[51] Thomson, Elf-Aquitaine, Bull, and BSN embarked on Europewide consultation procedures, facilitating the development of Euro–works councils. Not surprisingly, the first three belong to the public sector. This fit well with the industrial relations strategy of the Labor Ministry. Still, union representation remained low at smaller firms.

We need to be careful in projecting too far ahead. As longstanding institutions, unions do not succumb easily to obstacles. Interunion cooperation, ad hoc though it may be, takes place at the local and regional levels. There have been innumerable examples of cooperation for defensive action, much to the consternation of confederation leaders. Such cooperation extends to collective bargaining arrangements and more positive action. In line with the "Europe of regions," the ETUC has facilitated working groups within transnational regions, and the CGT has participated in these less visible forums.

By early 1994, the divorce between union and party seemed to be reaching finality in the CGT, thereby removing one of the key sources of division within the French labor movement. CGT leaders took public positions against joint membership in the PCF. Perhaps more important, the confederation was engulfed in debate over the very nature of unionism. In other words, what role should the traditional form of oppositional protest play in unions' strategic thinking? The organization has not yet solved the riddle, but the very debate suggests that institutions can adapt.

All the major French unions in the early 1990s had a new generation of leaders. To many, the oppositionalism of the past was only a historical relic, while propositionalism seemed utopian. Institutional memories remained powerful, but they were refracted through contemporary problems. *Union* commentators even suggested that the very *model* of unionism needed to be reconceptualized in light of changes taking place at work and in Europe.

Thus, adaptation will take place, even though it seems painfully slow to many observers. It will undoubtedly transform the old model of protest. Unions have too many possibilities to influence social policy to mobilize solely on the basis of a myth. This does not eliminate the utility of protest, but it means that protest must be mixed with other tools to influence decision making. The new model of labor politics will depend on the inclusiveness of the adjustment process.

Conclusion

Modernizers in the 1940s believed that state power could be mobilized for purposes that were politically determined, not economically constrained. Political power—if wielded with sufficient verve—could alter the allocation of economic resources, shift priorities, and change society. The modernizers succeeded in crafting a powerful state apparatus. They facilitated tremendous growth in the economy. They were less successful in changing social attitudes. In particular, the integration of labor into the postwar model was delayed until the 1980s, by which time the forces of decay in the union movement made such inclusion tenuous at best.

Political leaders two generations later reversed that ordering: economic constraints gained supremacy over political will. Policy making in the early 1990s was characterized by a fatalistic view that there was little margin for maneuver to spur employment and encourage economic activity. When growth sputtered in the early 1990s, for instance, both Left and Right governments insisted on maintaining budgetary stringency and a strong franc in spite of soaring unemployment. Policy makers decided to follow the German lead of financial orthodoxy and fiscal restraint rather than stimulating the domestic economy and risking exclusion from liberal Europe.

The combination of a strong state and excluded labor depended on a set of political coalitions that by the 1980s were no longer intact. In the 1940s and 1950s, the modernizers convinced "third force" governments of the necessity to develop interventionist tools. By the 1980s, however, only the Left believed in state voluntarism. The U-turn and the embrace of Europe effectively negated such thinking. Only the weakened Communists continued to believe in the state's capacity to mold society.

France's European policy helped transform the French model. Ironically, the modernizers envisioned a strong presence within Europe and voluntaristic policy instruments. They saw both as complementary. Instead, they were contradictory because the premise of the former was expansion of the firm, while the premise of the latter was the national state.

The French state, however, did not liquidate its policy instruments. Rather, they were used for new purposes. The French state became an instrument to construct market society—not unlike similar experiments across the Channel. French companies became internationally competitive, while French unions had difficulty organizing new categories of workers. At least part of French society—business—could no longer be considered weak.

The debilitation of labor organizations exacerbated the increasing polarization of labor markets, the durability of high unemployment, the poor quality of new jobs, the restlessness of early-retired workers, and racial tensions. The mixed response to Europe evident in the debates over Maastricht marked concern about the very nature of economic distribution, social interaction, and human community that have been challenged by market society and the European project.

What will replace the "French model" is open to question. French unions have historically relied on a combination of local militancy and state intervention to compensate for organizational inadequacies. In the 1990s, they can count on neither to serve their interests. Likewise, even those French firms strong at the European level display a reluctance to part company with the state. Parisian ministries protect industries from unfriendly competition to the east and south. They can also disburse investment funds for new research, finance social plans, and provide the framework for the development of human capital. Finally, governments remain responsive to protest reflexes—if they are sufficiently vocal.

While the boundaries of the national model have shifted, new patterns of adjustment will continue to integrate social interests and market constraints. Far from eliminating politics, adjustment in France will be forged by actors seeking to alter the distributional consequences of market competition.

Notes
Bibliography
Index

Notes

Chapter 1. Industrial Politics in France

1. See Henry W. Ehrmann, *Organized Business in France* (Princeton: Princeton University Press, 1957).

2. See John H. McArthur and Bruce R. Scott, *Industrial Planning in France* (Boston: Harvard Graduate School of Business, 1969).

3. See John Zysman, *Governments, Markets, and Growth: Financial Systems and the Politics of Industrial Change* (Ithaca: Cornell University Press, 1983), 157–60.

4. Suzanne Berger, "Lame Ducks and National Champions: Industrial Policy in the Fifth Republic," in *The Fifth Republic at Twenty*, ed. William G. Andrews and Stanley Hoffmann (Albany: State University of New York Press, 1981), 292–310.

5. A number of notable exceptions can be cited in the broader field of comparative political economy. See Michael J. Piore and Charles F. Sabel, *The Second Industrial Divide: Possibilities for Prosperity* (New York: Basic, 1984); Sabel, *Work and Politics: The Division of Labor in Society* (Cambridge: Cambridge University Press, 1982); Wolfgang Streeck, *Social Institutions and Economic Performance: Industrial Relations in Advanced Capitalist Economies* (Beverly Hills and London: Sage, 1992).

6. See John Zysman, "The French State in the International Economy," in *Between Power and Plenty: Foreign Economic Policies of Advanced Industrial States*, ed. Peter Katzenstein (Madison: University of Wisconsin Press, 1978).

7. For a representative example, see Mark Kesselman with Guy Groux, eds., *The French Workers' Movement: Economic Crisis and Political Change* (London: George Allen & Unwin, 1984). For a comparative approach, see Robert J. Flanagan, David W. Soskice, and Lloyd Ulman, *Unionism, Economic Stabilization, and Incomes Policies: European Experience* (Washington, D.C.: Brookings, 1983).

8. William James Adams, *Restructuring the French Economy: Government*

and the Rise of Market Competition Since World War II (Washington, D.C.: Brookings, 1989).

9. See Vivien Schmidt, *Democratizing France: The Political and Administrative History of Decentralization* (Cambridge: Cambridge University Press, 1991).

10. Alexis de Tocqueville, *The Old Régime and the French Revolution*, trans. Stuart Gilbert (New York: Anchor, 1856, 1955).

11. Michel Crozier, *The Bureaucratic Phenomenon* (Chicago: University of Chicago Press, 1964).

12. Michel Crozier, *La société bloquée* (Paris: Seuil, 1970).

13. See Stanley Hoffmann, "Paradoxes of the French Political Community," in *In Search of France: The Economy, Society, and Political System in the Twentieth Century*, ed. Hoffmann et al. (New York: Harper, 1963).

14. Stanley Hoffmann, "The Ruled: Protest as a National Way of Life," *Decline or Renewal? France Since the 1930s* (New York: Viking, 1974).

15. See William Schonfield, *Obedience and Revolt: French Behavior Toward Authority* (Beverly Hills and London: Sage, 1976); Philip G. Cerny, ed., *Social Movements and Protest in France* (New York: St. Martin's, 1982).

16. John S. Ambler, "Educational Pluralism in the French Fifth Republic," in *Searching for the New France*, ed. George Ross and James F. Hollifield (New York and London: Routledge, 1991), 193–220.

17. Suzanne Berger, *Peasants Against Politics: Rural Organization in Brittany, 1911–1967* (Cambridge: Harvard University Press, 1972); John T. S. Keeler, *The Politics of Neo-Corporatism in France* (New York and Oxford: Oxford University Press, 1987).

18. Robert Salisbury, "An Exchange Theory of Interest Groups," in *Interests and Institutions: Substance and Structure in American Politics* (Pittsburgh: University of Pittsburgh Press, 1992), 3–34.

19. Ezra N. Suleiman has exhorted social scientists to go beyond sensationalistic studies of alcohol lobbies, veterans groups, and partisans of Catholic education to examine systematically the relationship between interests and the French state. See *Politics, Power, and Bureaucracy in France: The Administrative Elite* (Princeton: Princeton University Press, 1974), 321–22.

20. Robert A. Dahl, *Democracy and Its Critics* (New Haven: Yale University Press, 1989).

21. Andrew Shonfield, *Modern Capitalism: The Changing Balance of Public and Private Power* (New York: Oxford University Press, 1969), 129.

22. It is not surprising that much of the Marxist writing on the "autonomy" of the capitalist state found its origin in France. See the work of Nicos Poulantzas, *State, Power, Socialism* (London: Verso, 1980).

23. Frank L. Wilson, "Alternative Models of Interest Intermediation: The Case of France," *British Journal of Political Science* 12, no. 2 (April 1982):

173–200; "French Interest Group Politics: Pluralist or Neocorporatist?" *American Political Science Review* 77, no. 4 (1983): 895–910; and "Les groupes d'intérêt sous la Cinquième République," *Revue française de science politique* 33, no. 2 (April 1983): 220–54.

24. Wilson, "French Interest Group Politics," 907.

25. John T. S. Keeler, "Situating France on the Pluralism-Corporatist Continuum: A Critique and Alternative to the Wilson Perspective," *Comparative Politics* 17, no. 2 (January 1985): 229–49.

26. Ibid., 244.

27. Keeler, *The Politics of Neo-Corporatism in France.*

28. Andrew Cox and Jack Hayward, "The Inapplicability of the Corporatist Model in Britain and France: The Case of Labor," *International Political Science Review* 4, no. 2 (1983): 217–40.

29. Pierre Birnbaum, *La Logique de l'état* (Paris: Fayard, 1982), 109.

30. Pierre Birnbaum, "The State versus Corporatism," *Politics and Society* 11, no. 4 (1982): 501.

31. Nelson W. Polsby, *Community Power and Political Theory: A Further Look at Problems of Evidence and Inference*, 2d ed. (New Haven: Yale University Press, 1980).

32. Dénis Segrestin, *Le Phénomène corporatiste: Essai sur l'avenir des systèmes professionnels fermés en France* (Paris: Fayard, 1985).

33. For comparative analyses of group-state relations, see Philippe C. Schmitter and Gerhard Lehmbruch, eds., *Trends Toward Corporatist Intermediation* (Beverly Hills and London: Sage, 1979); Lehmbruch and Schmitter, eds., *Patterns of Corporatist Policy-Making* (Beverly Hills and London: Sage, 1982).

34. Alan Cawson, ed., *Organized Interests and the State: Studies in Meso-Corporatism* (London and Beverly Hills: Sage, 1985).

35. See Frank Baumgartner, *Conflict and Rhetoric in French Policy Making* (Pittsburgh: University of Pittsburgh Press, 1989).

36. E. E. Schattschneider, *The Semisoveriegn People: A Realist's View of Democracy in America* (New York: Holt, Rinehart and Winston, 1960), 71.

37. Val R. Lorwin, *The French Labor Movement* (Cambridge: Harvard University Press, 1954).

38. Jane Jenson and George Ross, "The Tragedy of the French Left," *New Left Review* 171 (September/October 1988): 5–46.

39. Charles Tilly, *The Contentious French* (Cambridge: Harvard University Press, 1986).

40. Charles Lewis Taylor and David A. Jodice, *World Handbook of Political and Social Indicators*, vol. 2: *Political Protest and Government Change* (New Haven: Yale University Press, 1983).

41. We can quibble with the indicators or the relevance of the time period, but the evidence suggests that France is not particularly more protest-prone than other European democracies.

42. "Lacking an 'inside track' to the state, the [French] working class responds by activating an alternative vehicle of collective pressure, the strike." See Walter Korpi and Michael Shalev, "Strikes, Industrial Relations and Class Conflict in Capitalist Societies," *British Journal of Sociology* 30, no. 2 (June 1979): 182.

43. Herbert P. Kitschelt, "Political Opportunity Structures and Political Protest: Anti-Nuclear Movements in Four Democracies," *British Journal of Political Science* 16 (January 1986): 57–85.

44. See Leon N. Lindberg, John L. Campbell, and J. Rogers Hollingsworth, eds., *Governance of the American Economy* (Cambridge: Cambridge University Press, 1991).

45. According to a recent study, "sectors are one of the two nexuses—the other being the community or locality—where exchanges among producers and among producers, suppliers, and consumers can be socially constructed." They also "provide one of the principle frameworks within which public policies can be effectively administered—the other being the issue arena." J. Rogers Hollingsworth, Philippe C. Schmitter, and Wolfgang Streeck, "Capitalism, Sectors, Institutions, and Performance," in *Governing Capitalist Economic: Performance and Control of Economic Sectors*, ed. Hollingsworth, Schmitter, and Streeck (New York: Oxford, 1994), 9.

46. See Anthony M. Orum, Joe R. Feagin, and Gideon Sjoberg, "Introduction: The Nature of Case Studies," in *A Case for the Case Study*, ed. Feagin, Orum, and Sjoberg (Chapel Hill: University of North Carolina Press, 1991).

47. Orum, Feagin, and Sjoberg, eds., *A Case for the Case Study.*

Chapter 2. The Adjustment Triangle

1. François Caron, *An Economic History of Modern France*, trans. Barbara Bray (New York: Columbia University Press, 1979); Jean-Jacques Carré, Paul Dubois, and Edmond Malinvaud, *La croissance française: Un essai d'analyse économique causale de l'après-guerre* (Paris: Seuil, 1972); Stephen S. Cohen, *Modern Capitalist Planning: The French Model*, 2d ed. (Berkeley and Los Angeles: University of California Press, 1977); André Gauron, *Histoire économique et sociale de la Cinquième République: Le temps des modernistes*, vol. 1 (Paris: La Découverte, 1983) and *Histoire économique et sociale de la Cinquième République: Années de rêves, années de crises (1970–1981)*, vol. 2 (Paris: La Découverte/Maspero, 1988); John H. McArthur and Bruce R. Scott, *Industrial Planning in France* (Boston: Harvard Graduate School of Business, 1969); John Sheahan, *Promotion and Control of Industry in Postwar France* (Cambridge: Harvard University Press, 1963).

2. This discussion borrows heavily from Robert O. Paxton, *Vichy France: Old Guard and New Order, 1940–1944* (New York: Columbia University Press, 1972).

3. Richard F. Kuisel, *Capitalism and the State in Modern France* (Cambridge: Cambridge University Press, 1981), 155–56.

4. Stephen S. Cohen, *Modern Capitalist Planning.*

5. See Irwin W. Wall, "Jean Monnet, the United State and the French Economic Plan," in *Jean Monnet: The Path to European Unity*, ed. Douglas Brinkley and Clifford Hackett (New York: St. Martin's, 1991).

6. Philippe Mioche, *Le Plan Monnet: Genèse et élaboration, 1941–1947* (Paris: Publications de la Sorbonne, 1987), 92ff.

7. Jean Monnet, *Memoirs*, trans. Richard Mayne (London: Collins, 1978).

8. Richard F. Kuisel, "L'American Way of Life et les Missions Françaises de Productivité," *Vingtième siècle* 17 (January–March 1988): 21–38.

9. Michael Loriaux, *France After Hegemony: International Change and Financial Reform* (Ithaca: Cornell University Press, 1991).

10. See Ezra N. Suleiman, *Elites in French Society: The Politics of Survival* (Princeton: Princeton University Press, 1978).

11. Darryl Holter, *The Battle for Coal: Miners and the Politics of Nationalization, 1940–1950* (DeKalb: Northern Illinois University Press, 1992).

12. Philippe Mioche, "Le patronat de la sidérurgie française et le Plan Schuman en 1950–1952: Les apparences d'un combat et la réalité d'une mutation," in *The Beginnings of the Schuman-Plan*, ed. Klaus Schwabe (Baden-Baden: Nomos Verlag, 1988), 305–18.

13. See Erhard Friedberg, "Administration et entreprises," in *Où va l'administration française?*, ed. Michel Crozier (Paris: Éditions d'Organisation, 1974), 101–40, and Bruno Jobert, "Le Ministère de l'Industrie et la cohérence de la politique industrielle, *Revue française de science politique* 23, no. 2 (April 1973): 321–29.

14. Frances Lynch, "The Role of Jean Monnet in Setting up the European Coal and Steel Community," in *The Beginnings of the Schuman-Plan*, ed. Schwabe, 118.

15. McArthur and Scott, *Industrial Planning in France*, 330.

16. Jack Hayward argues that intense rivalry characterizes the French policy community. Our account does not dispute this claim but rather suggests that the French state has been able to mobilize resources to influence industrial outcomes. See *The State and the Market Economy: Industrial Patriotism and Economic Intervention in France* (New York: New York University Press, 1986), 19–38.

17. Richard F. Kuisel, "L'American Way of Life," 29–30.

18. Mioche, *Le Plan Monnet*, 95.

19. Cohen, *Modern Capitalist Planning*, xv.

20. Ezra N. Suleiman, "From Right to Left: Bureaucracy and Politics in France," in *Bureaucrats and Policy Making: A Comparative Overview*, ed. Suleiman (New York: Holmes & Meier, 1984).

21. See John T. S. Keeler, "Corporatist Decentralization and Commercial Modernization in France: The Royer Law's Impact on Shopkeepers, Supermarkets and the State," in *Socialism, the State and Public Policy in France*, ed. Philip G. Cerny and Martin A. Schain (New York: Methuen, 1985), 265–91.

22. Élie Cohen, *L'État brancardier: Politiques du déclin industriel (1974–1984)* (Paris: Calmann-Lévy, 1989).

23. This discussion of the Trésor depends heavily on an account in John Zysman, *Governments, Markets, and Growth: Financial Systems and the Politics of Industrial Change* (Ithaca: Cornell University Press, 1983), 99–169.

24. The Conservative government of Jacques Chirac privatized Paribas and the Société Générale in 1987.

25. See the Gerschenkron-Landes debate on economic structure and cultural attitudes. David S. Landes, "French Entrepreneuship and Industrial Growth in the Nineteenth Century," *Journal of Economic History*, May 1949, 45–61. For the critique of "backwardness," see Alexander Gerschenkron, "Social Attitudes, Entrepreneurship, and Economic Development," in *Economic Backwardness in Historical Perspective* (Cambridge: Harvard University Press, 1962), 52–71.

26. Maurice Lévy-Leboyer, "Le patronat français a-t-il été malthusien?" *Mouvement social* 88 (July–September 1974): 3–49.

27. Zysman, *Governments, Markets, and Growth*, 122.

28. Peter Hall, *Governing the Economy: The Politics of State Intervention in Britain and France* (New York: Oxford University Press, 1986), 154.

29. Andrew Shonfield, *Modern Capitalism: The Changing Balance of Public and Private Power* (London and New York: Oxford University Press, 1965, 1969), 166.

30. Hall, *Governing the Economy*, 155–63.

31. Cohen, *Modern Capitalist Planning*, 56.

32. Vivien A. Schmidt, "Through 'More State' and 'Less State,' the French State Goes On: Patterns of Industrial Policy-Making in France 1981–1991," presented at the annual meeting of the American Political Science Association, Washington, D.C., 1991.

33. Vivien A. Schmidt, "An End to French Economic Exceptionalism? The Transformation of Business under Mitterrand," in *The Mitterrand Era: Policy Alternatives and Political Mobilization in France*, ed. Anthony Daley (Basingstoke and New York: Macmillan and New York University Press, 1995).

34. John Zysman, "Inflation and the The Politics of Supply," in *The Politics of Inflation*, ed. Leon Lindberg and Charles S. Maier (Washington, D.C.: Brookings Institution, 1985).

35. François Bloch-Lainé, *Pour une réforme de l'entreprise* (Paris: Seuil, 1963), 45.

36. George Ross, "The Perils of Politics: French Unions and the Crisis of

the 1970s," in *Unions, Change and Crisis: French and Italian Union Strategy and the Political Economy, 1945–1980*, ed. Peter Lange, George Ross, and Maurizio Vannicelli (London and Boston: George Allen & Unwin, 1982), 71.

37. Chris Howell and Anthony Daley, "Introduction: The Transformation of Political Exchange," *International Journal of Political Economy* 23, no. 4 (Winter 1992–93).

38. Jacques Freyssinet, *Le chômage* (Paris: La Découverte, Collection Repères, 1987); Jean-Michel Fahy, *Le chômage en France*, 4th ed. (Paris: P.U.F., Que sais-je?, 1989).

39. See Thierry Beranger, "1959–1974: quinze ans d'assurance chômage," *Droit social*, September–October 1975, 135–141.

40. Compare with Britain and West Germany, for instance, where employers and employees split the payroll tax.

41. Jean-Paul Géhin, "Les institutions de formation continue: Diversité des interventions patronales," *Formation emploi* 11 (July–September 1985): 48–61.

42. See Jean-Yves Hocquet, "Les conventions du Fonds national de l'emploi," *Droit social*, September–October 1985, 651–57.

43. See Nicole Kerschen and Anne-Valérie Nenot, "Délégation à l'emploi et négociation des conventions du Fonds national de l'emploi: la pratique des contreparties," *Droit social*, January 1989, 17–22.

44. Cour des Comptes, *Rapport au Président de la République*, 1989, 215–22.

45. Suzanne Berger, "The Traditional Sector in France and Italy," in Suzanne Berger and Michael J. Piore, *Dualism and Discontinuity in Industrial Societies* (Cambridge: Cambridge University Press, 1980).

46. The estimate is from George Ross, "The Confédération Générale du Travail in Eurocommunism," *Politics and Society* 9, no. 1 (1979): 40–41, n. 19.

47. See Ross, *Workers and Communists*.

48. Compare, for instance, articles in the two main organs of union and party, *Le Peuple* (CGT) and *L'Humanité* (PCF).

49. See Alain Bergounioux, *Force ouvrière* (Paris: Seuil, 1975).

50. Gérard Adam, Frédéric Bon, Jean Capdevielle, and René Mouriaux, *L'ouvrier français en 1970* (Paris: A. Colin, 1971).

51. See Michel Branciard, *Syndicats et partis: Autonomie ou dépendance, 1948–1981* (Paris: Syros, 1982), 2:36. See also Bergounioux, *Force ouvrière*.

52. For the tension between the minority faction of the CFTC, which was later to gain control and create the CFDT, and the MRP, see Paul Vignaux, *De la CFTC à la CFDT: Syndicalisme et socialisme, "Reconstruction" (1946–1972)* (Paris: Éditions ouvrières, 1980). See also Branciard, *Syndicats et partis*, 45–66.

53. See Hervé Hamon and Patrick Rotman, *La deuxième gauche: Histoire intellectuelle et politique de la CFDT* (Paris: Éditions Ramsay, 1982).

54. See Gérard Adam, *Le pouvoir syndical* (Paris: Dunod, 1983); and Hubert Landier, *Demain, quels syndicats?* (Paris: Livre de Poche, 1981).

55. Jean-Pierre Duprilot and Paul Fieschi-Vivet, *Les comités d'entreprise (Droit et pratique)* (Paris: P.U.F.-Que sais-je?, 1982).

56. Works councils have been mandated since 1945. In large plants, they handle budgets sometimes of several million dollars which are disbursed for a functions ranging from summer camp for workers' children to subsidies for the company canteen to Christmas presents. A collective bargaining agreement in 1969 required managers to consult works councils on redundancies. An "interprofessional" accord in 1974 required a "social plan" over the terms of those redundancies. The Auroux Laws in the early 1980s enabled works councils to employ an independent "expert" to judge the competitiveness of the firm. Yet the ability of the labor movement by then to wield these new instruments had been so reduced as to render them almost ineffective.

57. These social security elections have not been continuous. Elections were held after the war—in 1947, 1950, 1955, and 1962. Then the Gaullists suspended electoral choice of worker members of these boards, preferring instead to nominate employee representatives. The Socialists reinstated the elections in 1982, although, as of 1995, they have not been repeated.

58. Despite the preponderance of confederal votes, abstentions (ranging around 25 percent) should not be overlooked. The percentage of nonunion candidate crept upward after 1981 to constitute the largest vote "total" by the early 1990s.

59. The battle over representativeness can be followed in the union press. An article in the CFTC *Voix des métaux* 25 (April 1948) complained that the CGT tried to bar the CFTC as well as the FO from competing in elections in the Micheville plant. The CGT's version of the distortion of the concept of representativeness can be found in the metalworkers' *Voix des métallurgistes.*

60. See Robert J. Flanagan, David W. Soskice, and Lloyd Ulman, *Unionism, Economic Stabilization, and Incomes Policies: European Experience* (Washington, D.C.: Brookings, 1983), 567–648.

61. Michelle Durand, "De l'analyse causale à l'analyse structurelle, la négociabilité des conflits," *Sociologie du travail* 19, no. 4 (October–December 1977); "La grève: conflit structurel, système de relations industrielles ou facteur de changement social," *Sociologie du travail* 21, no. 3 (July–September 1979).

62. Georges Lefranc, *Le Mouvement syndical en France: De la Libération aux événements de mai–juin 1968* (Paris: PUF, 1969); Bernard H. Moss, *The Origins of the French Labor Movement, 1830–1914: The Socialism of Skilled Workers* (Berkeley and Los Angeles: University of California Press, 1976).

63. Flanagan, Soskice, and Ulman, *Unionism, Economic Stabilization, and Incomes Policies*, 576.

64. Anthony Daley, "Remembrance of Things Past: The Union-Party Linkage in France," *International Journal of Political Economy* 23, no. 4 (Winter 1992–93): 53–71.

65. "Days of action" are not only organized for party political purposes. They also help forge temporary unity in plants or regions over a set of labor market (collective bargaining) demands.

66. This section is inspired by work on "Fordism." See Michael J. Piore and Charles F. Sabel, *The Second Industrial Divide: Possibilities for Prosperity* (New York: Basic, 1984). See also the voluminous writings of the French Regulation School. For instance, see Robert Boyer, *La Théorie de la régulation: Une analyse critique* (Paris: La Découverte, 1987) and "Wage Labor, Capital Accumulation, and the Crisis, 1968–82," in *The French Workers' Movement: Economic Crisis and Political Change*, ed. Mark Kesselman (London: Allen & Unwin, 1984). For a synthetic approach, see Robert Brenner and Mark Glick, "The Regulation Approach: Theory and History," *New Left Review* 188 (July–August 1991).

67. Kuisel, *Capitalism and the State in Modern France*, 248.

68. Pierre Rosanvallon, "The Development of Keynesianism in France," in *The Political Power of Economic Ideas: Keynesianism Across Nations*, ed. Peter A. Hall (Princeton: Princeton University Press, 1989), 171–93.

69. See Chris Howell, *Regulating Labor: The State and Industrial Relations Reform in Postwar France* (Princeton: Princeton University Press, 1992).

70. Carré, Dubois, and Malinvaud, *La croissance française*.

71. Henri Weber, *Le parti des patrons: Le CNPF (1946–1986)* (Paris: Seuil, 1986); Henry W. Ehrmann, *Organized Business in France* (Princeton: Princeton University Press, 1957).

72. Benjamin Coriat, *L'Atelier et le chronomètre: Essai sur le Taylorisme, le Fordisme et la production de masse*, 2d ed. (Paris: C. Bourgois, 1982).

73. Gauron, *Histoire économique et sociale*, 1:114; Bloch-Lainé, *Pour une reforme de l'entreprise*.

74. Dean Savage, *Founders, Heirs, and Managers: French Industrial Leadership in Transition* (Beverly Hills and London: Sage, 1979).

75. See Maurice Lévy-Leboyer, "The Large Corporation in Modern France," in *Managerial Hierarchies: Comparative Perspectives on the Rise of the Modern Industrial Enterprise*, ed. Alfred Chandler (Cambridge: Havard University Press, 1980), 117–60.

76. Gauron, *Histoire économique et sociale*, vol. 2.

77. Alain Lipietz, "Governing the Economy in the Face of International Challenge: From National Developmentalism to National Crisis," in *Searching*

for the New France , ed. James F. Hollifield and George Ross (New York: Routledge, 1991), 25.

Chapter 3. The Postwar Steel Industry

1. Raymond Cartier, "La Lorraine, Texas français," *Paris-Match*, June 26, 1961.

2. For a much fuller explanation of steelmaking and industry structure, see U.S. Congress, Office of Technology Assessment, *Technology and Steel Industry Competitiveness* (Washington, D.C.: USGPO, 1980); Donald Barnett and Louise Schorsch, *Steel: Upheaval in a Basic Industry* (Cambridge: Ballinger, 1983); William T. Hogan, *World Steel in the 1980s: A Case of Survival* (Lexington, Mass.: D.C. Heath, 1983).

3. Philippe Saint-Marc, *La France dans la C.E.C.A.: Une expérience de planifications multiples du charbon et de l'acier*, Cahier de la Fondation Nationale des Sciences Politiques, no. 114 (Paris: Armand Colin, 1961), 248–49.

4. The canal failed to live up to expectations. Backers promised that it would bring in German coal and take out Lorraine iron ore for use in Germany. However, German producers stopped making Thomas steel in the late 1950s. Later, the waterway could not accommodate the superbarges used to transport raw materials worldwide.

5. Saint-Marc, *La France dans la C.E.C.A.*, 252. See also Groupement de l'Industrie Sidérurgique (GIS), *Rapports annuels*, various years.

6. Jean-Noël Jeanneney, *François de Wendel en République: L'argent et le pouvoir* (Paris: Seuil, 1976).

7. Richard F. Kuisel, *Capitalism and the State in Modern France: Renovation and Economic Management in the Twentieth Century* (Cambridge: Cambridge University Press, 1981), 207–11.

8. The future president of the CSSF Jacques Ferry claimed that nationalization remained a veiled threat to the industry for the entire postwar period. See his book-length interview in Philippe Mioche, *Jacques Ferry et la sidérurgie française depuis la Seconde Guerre Mondial* (Aix-en-Provence: Publications de l'Université de Provence, 1993).

9. See Philippe Mioche, *Le plan Monnet: Genèse et élaboration, 1941–1947* (Paris: Publications de la Sorbonne, 1987), 249–58; Philippe Mioche and Jacques Roux, *Henri Malcor: Un héritier des maîtres de forge* (Paris: Editions du CNRS, 1988), 188–89.

10. Henry W. Ehrmann, "The French Trade Associations and the Ratification of the Schuman Plan," *World Politics* 6, no. 4 (July 1954): 453–81; for a revisionist account that argues that the producers had serious market concerns, see Philippe Mioche, "Le patronat de la sidérurgie française et le Plan Schuman en 1950–1952: Les apparences d'un combat et la réalité d'une mutation," in

The Beginnings of the Schuman-Plan, ed. Klaus Schwabe (Baden-Baden: Nomos Verlag, 1988), 305–18.

11. Frances Lynch, "The Role of Jean Monnet in Setting up the European Coal and Steel Community," in *The Beginnings of the Schuman-Plan*, ed. Schwabe, 126; François Roth, "Les milieux sidérurgiques et l'annonce du Plan Schuman," in *The Beginnings of the Schuman-Plan*, ed. Schwabe, 367–80.

12. John Gillingham, *Coal, Steel, and the Rebirth of Europe, 1945–1955: The Germans and French from Ruhr Conflict to Economic Community* (Cambridge: Cambridge University Press, 1991).

13. Frances M. B. Lynch, "Resolving the Paradox of the Monnet Plan: National and International Planning in French Reconstruction," *Economic History Review*, 2d ser. 37, no. 2 (May 1984): 229–43.

14. Patrick A. Messerlin, "The European Iron and Steel Industry and the World Crisis," in *The Politics of Steel: Western Europe and the Steel Industry in the Crisis Years (1974–1984)*, ed. Yves Mény and Vincent Wright (Berlin and New York: de Gruyter, 1987), 126.

15. Gillingham, *Coal, Steel and the Rebirth of Europe*, 313ff.

16. Louis Lister, *Europe's Coal and Steel Community: An Experiment in Economic Union* (New York: Twentieth Century Fund, 1960).

17. William Diebold, Jr., *The Schuman Plan: A Study in Economic Cooperation* (New York: Praeger, 1959), 320–49.

18. Claude Prêcheur, *La Lorraine sidérurgique* (Paris: SABRI, 1959), 300.

19. Hayrettin Erdemli, Bernard Réal, Pierre Judet, and C. Palloix, *L'Internationalisation dans la sidérurgie* (Grenoble: Université des sciences sociales, Institut de Recherche, Economique et de Planification, 1974).

20. Labor productivity data must be handled with care. For instance, German figures are biased upward because steelmakers outsource peripheral activities, such as coke making, and they make a smaller number of finished products. See Pierre Gonod, *Problèmes et méthodes de mesure de la productivité dans les industries de la Communauté* (Paris: Dalloz and Sirey, 1961).

21. Bernard Lalanne, "Mort d'un Lobby," *L'Expansion*, June 1979, 108.

22. See Jean-Louis Masson, *L'Industrie lorraine du fer . . . ou comment tuer la poule aux oeufs d'or* (Paris: Cujas, 1977), 196ff.

23. Jean-Pierre Marger, *La situation et l'avenir de la sidérurgie*, Report to the Conseil Économique et Social, September 28, 1979, 15.

24. See Michel Freysennet, *La sidérurgie française, 1945–1979: L'Histoire d'une faillite, Les solutions qui s'affrontent* (Paris: Savelli, 1979); Benjamin Stora, *Crise, Puissance, Perspectives de la Sidérurgie Mondiale* (Paris: Economica, 1979).

25. Jack Hayward, *The State and the Market Economy: Industrial Patriotism and Economic Intervention in France* (New York: New York University Press, 1986), 74–75.

26. Andrew Shonfield, *Modern Capitalism: The Changing Balance of Public and Private Power* (London and New York: Oxford University Press, 1965, 1969), 138–40.

27. Dirk Spierenburg and Raymond Poidevin, *Histoire de la Haute Autorité de la Communauté du Charbon et de l'Acier: Un expérience supranationale* (Brussels: Bruylont, 1993), 580–87, 700–01.

28. See Chambre Syndicale de la Sidérurgie, "La verité sur la sidérurgie," Paris, September 1978.

29. See Saint-Marc, *La France dans la C.E.C.A.*, 248.

30. Mioche, *Jacques Ferry*, 167–68.

31. John H. McArthur and Bruce R. Scott, *Industrial Planning in France* (Boston: Harvard Graduate School of Business, 1969), 198.

32. Ibid., 201.

33. The Comptoir des Produits Sidérurgiques served as a coordinating body to enforce market discipline and to uphold the French end of the International Steel Cartel in the prewar period. The ECSC outlawed such behavior but the office continued to keep track of sales and garner statistics.

34. Donald Reid, "The Limits of Paternalism: Immigrant Coal Miners' Communities in France, 1919–45," *European History Quarterly* 15, no. 1 (January 1985), 99.

35. See, for instance, Daniel J. Walkowitz, *Worker City, Company Town: Iron and Cotton-Worker Protest in Troy and Cohoes, New York, 1855–84* (Urbana and Chicago: University of Illinois Press, 1978).

36. Archives of the Department of Moselle (ADM), *Fonds de cabinet (1944–1950)*, no. 191, Renseignements Généraux de Thionville, no. 855/50, December 6, 1950.

37. Chambre Syndicale de la Sidérurgie Française, "La sidérurgie française et le problème du logement," Paris, n.d. [1955 or 1956].

38. When the occupied territories returned to France in 1919, the secular Ferry Laws of 1884 were not applied. Instead, public schooling was governing by the Loi Falloux from 1850 and from German law during the occupation. See Lucie Tanguy, *Le capital, les travailleurs et l'école: L'exemple de la Lorraine sidérurgique* (Paris: Maspero, 1976). Compulsory Catholic education partially explains the relative conservatism of Moselle workers in comparison to their colleagues in the Meurthe-et-Moselle.

39. Eugène Descamps, *Militer: Une vie pour un engagement collectif* (Paris: Fayard, 1971), 53. The memoirs of Descamps offer insights into the social history of the Lorraine steel regions in the 1950s.

40. René Carême, *Combats d'un militant* (Paris: Éditions Ouvrières, 1977), 39.

41. See the series of three articles by S. Simon, "Histoire d'une grève: La grève de Micheville (28 janvier–16 mars 1951)," *Droit social* 14, nos. 6–8 (June–October 1951).

42. Archives of the Department of Moselle (ADM), Fonds de cabinet (1944–1950), no. 191, Rapports de police, note d'information, Thionville no. 881, December 12, 1950.

43. *La vie ouvrière*, July 19–25, 1955, excerpted in *L'Homme du fer: Mineurs de fer et ouvriers sidérurgistes lorrains, 1930–1959*, ed. Serge Bonnet, with Etienne Kagan and Michel Maigret (Nancy: Centre Lorrain d'Études Sociologiques, 1977), 2:327–30. See also Pierre Belleville, "La sidérurgie lorraine et son prolétariat," *Les Temps modernes* 191 (April 1962): 1509.

44. The High Authority found that 30 percent of all workers in the French steel industry were non-native in 1962. The country with the second highest number of non-natives was Belgium with 18 percent. The Community average was 12 percent. See ECSC, High Authority, *General Report of the Activities of the Community*, May 1963, 658–59.

45. See Odette Hardy-Hémery, *De la croissance à la désindustrialisation: Un siècle dans le Valenciennois* (Paris: Presses de la Fondation Nationale des Sciences Politiques, 1984).

46. Andrée Michel, "L'Immigration algérienne en Moselle," *Annales de Géographie* 65, no. 5 (September–October 1956).

47. Ibid.

48. See ADM, Fonds de Cabinet (1944–1950), no. 191, Renseignements Généraux de Thionville, no. 843, November 29, 1950. The issue of purchasing power can be followed in local union publications. See *La Lorraine sociale*, November 1951, for a representative article.

Chapter 4. Labor's Response to Exclusion

1. Fernand Dupuy, *Etre maire communiste* (Paris: Calmann-Lévy, 1975), 106.

2. See Serge Bonnet, *Sociologie politique et religieuse de la Lorraine*, Cahiers de la Fondation Nationale des Sciences Politique, no. 181 (Paris: Armand Colin, 1972), 359–428.

3. Bonnet estimates that over 50 percent of the Communists in Meurthe-et-Moselle in 1945 were foreigners, most of them Italian. See Serge Bonnet, "Political Alignments and Religious Attitudes within the Italian Immigration to the Metallurgical Districts of Lorraine," *Journal of Modern History* 2, no. 2 (Winter 1968): 135.

4. Bonnet, "Political Alignments," 134–39.

5. Serge Bonnet's most elaborate treatment of "Italian communism" in Lorraine can be found in his doctoral thesis: "Les ouvriers migrants quotidiens des usines sidérurgiques de l'agglomération de Longwy, 1962–1963," Faculté des Lettres de Paris, 1965.

6. Gérard Noiriel, *Longwy: Immigrés et prolétaires, 1880–1980* (Paris: PUF, 1984), 365–66.

7. Even in the late 1980s differences were visible in interunion relations in Lorraine and in the Nord. The author was struck by the mutual contempt of unions during interviews in the Nord. In Lorraine, militants at least showed respect for their colleagues in the other union.

8. See editorial in the CFTC-FGM's *Voix des métaux*, November 1952, for a biting critique of both industry and state. It accused the owners of refusing to apply legislation on collective bargaining, of illegally laying off délégués, of denying union rights, of impeding the operations of works councils, of pursuing a policy of "class struggle." It accused the government of complicity for refusing to move against the industry.

9. For an analysis of the metamorphosis of the CFTC into the CFDT, see Paul Vignaux, *De la CFTC à la CFDT: Syndicalisme et socialisme, "Reconstruction" (1946–1972)* (Paris: Éditions ouvrières, 1980).

10. The events of 1958 and the near putsch constituted the major exception. Still, the CFTC rarely called days of actions in support of national political platforms.

11. See *Voix des métaux*, no. 3, January 1952.

12. ADM, Fonds de Cabinet (1944–1950), no. 191, Renseignements Généraux, Note d'information, no. 5909, n.d. See also Eugène Descamps, *Militer: Une vie pour un engagement collectif* (Paris: Fayard, 1971), 64–65.

13. The estimate is from Descamps, *Militer*, 65. A police report concerning the FO was somewhat condescending: "Objectively, it is important to remember that the FO represents little in Moselle metallurgy." ADM, Fonds de Cabinet (1944–1950), no. 191, Dossier Police, Renseigements Généraux de Thionville, no. 828, November 11, 1950.

14. George Ross argues that the strike calls were not made from the Confederation for tactical purposes. Still, the Confederation coordinated the industrial actions that developed in October and peaked in December 1947. See *Workers and Communists in France: From Popular Front to Eurocommunism* (Berkeley and Los Angeles: University of California Press, 1982), 51–58.

15. René Schwob, "Un Futur centenaire, Le syndicalisme français" (Metz: CGT-FO, 1981).

16. Noiriel, *Longwy: Immigrés et prolétaires*, 322.

17. *Paris-Press*, October 10–11, 1948, excerpted in *L'Homme du fer: Mineurs de fer et ourvriers sidérurgistes lorrains, 1930–1959*, ed. Serge Bonnet, with Etienne Kagan and Michel Maigret (Nancy: Centre Lorrain d'Études Sociologiques, 1977), 2:249–51.

18. For a joint communiqué issued by the CGT-FTM and CFTC-FGM, see *Voix des métaux*, April–May 1954.

19. The insistence on the Resistance program continued into the late 1950s. See *La vie ouvrière* during this period for representative examples.

20. CGT-FTM, "Rapports d'activité et perspectives d'organisation administratif et financier," 15e Congrès fédéral, April 4, 1946.

21. *Voix des métaux*, September 1952.

22. Ambroise Croizat, "Rapport d'activité," in CGT-FTM, 17e Congrès fédéral, May 31–June 3, 1950.

23. CGT-FTM, 18e Congrès fédéral, November 14–18, 1952.

24. Jean Breteau, "Rapport d'activité," in 18e Congrès fédéral.

25. *Voix des métaux*, January 1952.

26. See also *Lorrain social*, March 1952.

27. See *La métallurgie syndicaliste*, February 1952.

28. For a representative article, see *La métallurgie syndicaliste*, February 1951.

29. CGT-FTM, 19e Congrès fédéral, November 27–December 1, 1954, 194.

30. CGT-FTM, *Guide du métallurgiste*, January 1955.

31. See Ross, *Workers and Communists*, 49–82.

32. Noiriel, *Longwy: Immigrés et prolétaires*, 330.

33. *Le guide du métallurgiste, sidérurgie*, January 1955.

34. See *Le sidérurgiste lorrain*, January 1964.

35. See Descamps, *Militer; Vignaux, De la CFTC à la CFDT.*

36. Ross, *Workers and Communists*, 150.

37. The 1950 law—voted by the National Assembly February 11, 1950—set the framework for French industrial relations for the following twenty years. The law favored sectoral over company or national "interprofessional" agreements. It also specified in very elaborate terms the differences between types of collective bargaining agreements. For an early evaluation of the law, see the three-part series by Paul Durand, "La loi du 11 février 1950 sur les conventions collectives du travail," *Droit social* 13, no. 3 (March 1950): 93–101; no. 4 (April 1950): 155–63; no. 5 (May 1950): 186–93.

38. See, for instance, François Sellier, *La confrontation sociale en France, 1936–1981* (Paris: PUF, 1984); Jean-Daniel Reynaud, *Les syndicats en France* (Paris: Seuil, 1975); Gérard Adam, Reynaud, and Jean-Maurice Verdier, *La négociation collective en France* (Paris: Éditions Ouvrières, 1972).

39. The steel industry dominated the UIMM in ways similar to its influence in the peak employers' confederation Chambre Nationale du Patronat Français. In the CNPF, a member from the Chambre Syndicale de la Sidérurgie Française always played a major organizational role. For instance, Jacques Ferry was vice-president of the CNPF for over twenty years. In the UIMM, the CSSF played an even larger role due to the size of its employment vis-à-vis other metalworking industries.

40. François Sellier, "La cohésion patronale dans les négociations sociales: L'exemple de l'union des industries métallurgiques et minières," *Droit social* 23, nos. 7–8 (July–August 1960): 412. See also Sellier's follow-up article a decade later, "L'évolution des négociations collectives dans la sidérurgie et la métal-

lurgie en France (1950–1969)," *Droit social* 33, nos. 9–10 (September–October 1970).

41. Reluctance to engage in plant-level agreements continued to characterize employer attitudes into the 1970s. See Jean Bunel and Jean Saglio, "La faiblesse de la négociation collective et le pouvoir patronal," *Sociologie du travail* 19, no. 4 (October–December 1977).

42. See Val R. Lorwin, *The French Labor Movement* (Cambridge: Harvard University Press, 1954), 216.

43. See Michel Freyssenet, *La sidérurgie française, 1945–1979: L'historie d'une faillite. Les solutions qui s'affrontent* (Paris: Savelli, 1979).

44. Certain modes of behavior were more acceptable than others. For instance, the CFTC in Longwy complained in 1948 that the CGT was distributing free unions cards to its members along with "complimentary" fuel oil and bags of sugar (*La Lorraine sociale*, March–April 1948). It complained in 1951 when the CGT distributed fliers to all steelworkers urging them to support its demands and to forge a single union at the base (*La Lorraine sociale*, December 1951). Most of these complaints ended by the mid-1950s, indicating at least tacit acceptance by the CGT of a plurality of working-class representation.

45. FO publications during the 1950s and 1960s rarely missed the opportunity to refer to the CGT in terms relating to the PCF, the Komintern, the Korminform, or the division of the CGT between 1921 and 1936.

46. The demands for the the centralization of collective bargaining can be found in the CGT-FTM, *Vie ouvrière-métaux* or the CFTC/CFDT, *Voix des métaux*.

47. Industrial relations legislation permits release time for works council members and délégués. Still, union newspapers frequently recounted the resentment of other workers toward the union militant who was absent from his shift. Collective bargaining sessions increased the pressure on already overworked militants.

48. As the negotiating region of the Nord grew in industrial importance after the construction of the greenfield plant at Dunkerque, negotiations built on and sometimes even surpassed Lorraine agreements. Yet, this catch-up did not take place until the late 1960s.

49. See, for example, the autobiographical work by George Steines, *Le fer dans la peau* (Metz: Éditions Serpenoise, 1977).

50. See, for instance, John T. Dunlop, *Industrial Relations Systems* (New York: Henry Holt, 1958).

51. Chris Howell, *Regulating Labor: The State and Industrial Relations Reform in Postwar France* (Princeton: Princeton University Press, 1992).

Chapter 5. The Creation of National Champions in Steel

1. For a general discussion of national champions, see John Zysman, *Political Strategies for Industrial Order: State, Market, and Industry in France*

(Berkeley and Los Angeles: University of California Press, 1977), 62–63. For the intellectual argument on the importance of national champions, see Lionel Stoleru, *L'Impératif industriel* (Paris: Seuil, 1969).

2. Philippe Saint-Marc, *La France dans la C.E.C.A.: Une expérience de planifications multiples du charbon et de l'acier*, Cahier de la Fondation Nationale des Science Politiques, no. 114 (Paris: Armand Colin, 1961), 121.

3. The French steelmakers used overtime much more readily than other competitors in the ECSC. A Community study found that by 1962, overtime constituted over 28 percent of normal hours worked. The closest "competitor" was Luxemburg, whose steelworker overtime was 8 percent of normal hours. See ECSC, *High Authority, General Report of the Activities of the Community*, March 1964, 351. It should be remembered, however, that the legal work week was forty hours in France.

4. These figures and their interpretation come from an unpublished report excerpted in *L'Homme du fer: Mineurs de fer et ouvriers sidérurgistes lorrains 1960–1973*, ed. Serge Bonnet, with Etienne Kagan and Michel Maigret (Nancy and Metz: Presses Universitaires de Nancy and Éditions Serpenoise, 1984), 3:35–37.

5. Claude Durand, "L'Évolution du travail dans les laminoirs," *Revue française du travail* 13, no. 1 (January–March 1959), pp. 3–18.

6. David Charasse and Gérard Noiriel, "Marcel Donati: Ouvrier Lamineur" (interview), *Travail* 4 (April 1984), pp. 25–27.

7. Jacques Dofny, Claude Durand, Jean-Daniel Reynaud, and Alain Touraine, *Les ouvriers et le progrès technique. Étude de cas: Un nouveau laminoir* (Paris: Armand Colin, 1966).

8. For a representative collection of union complaints on work organization, see the CGT's *Le Guide du métallurgiste, sidérurgie*, no. 3 (January 1955), 14–45.

9. *Actualités Industrielles Lorraines*, no. 43 (May–June 1956).

10. Burkart Lutz and Alfred Willener, *Niveau de mécanisation et mode de rémuneration* (Luxemburg: European Coal and Steel Community, High Authority, May 1960), 50–64.

11. ECSC, *High Authority, General Report of the Activities of the Community*, March 1967, 29.

12. Jean Baumier, *La fin des maîtres de forges* (Paris: Plon, 1981), 85.

13. Giscard d'Estaing, who was finance minister in 1965, put this condition in writing. See Jean G. Padioleau, *Quand la France s'enferre: La politique sidérurgique de la France depuis 1945* (Paris: PUF, 1981), 52–53.

14. An earlier merger had taken place in 1963 to create the Société Mosellane de Sidérurgie (SMS). The two companies—the Union des Consommateurs de Produits Métallurgiques et Industriels (UCPMI) and the Société Métallurgique de Knutange (SMK)—together produced 1.7 MT of steel in 1962, the

total ranking fifth among the French producers. The UCPMI integrated plant in Hagondange, which produced heavy joists and merchant bars, had been built by Thyssen in 1912 and bought on favorable terms in 1920 from the French government after it expropriated German property in the liberated territories after World War I. The owners were a consortium of metalworking industries led by Renault, Peugeot, and Fives-Lille. The SMK plant was also German-made and confiscated in 1919. It produced an array of construction-grade products. The combined SMS also owned important iron mining concessions.

15. In the complicated ownership pattern in French steel, Wendel-Sidélor owned 65 percent of Sollac and 100 percent of Sacilor.

16. "Convention générale entre l'état et l'industrie sidérurgique," July 1966, 9. Scholars have noted as late as 1983 that the text was never made public. However, a copy can be found in the archives of the CFDT-FGMM.

17. The SMN was a small integrated steel plant owned by the Schneider group and located outside Caen in Normandy. Destroyed during the war, it was rebuilt after 1945 with a capacity to produce wire and merchant bars.

18. Jack Hayward, *The State and the Market Economy: Industrial Patriotism and Economic Intervention in France* (New York: New York University Press, 1986), 79–80.

19. Jack Hayward, "The Nemesis of Industrial Patriotism: The French Response to the Steel Crisis," in *The Politics of Steel: Western Europe and the Steel Industry in the Crisis Years (1974–1984)*, ed. Yves Mény and Vincent Wright (Berlin and New York: de Gruyter, 1987), 519.

20. Because plans were not finalized in 1964, the plant was not included in the Convention.

21. Padioleau, *Quand la France s'enferre*, 114.

22. Sidney Tarrow, "Regional Policy, Ideology, and Peripheral Defense: The Case of Fos-sur-Mer," in *Territorial Politics in Industrial Nations*, ed. Sidney Tarrow, Peter J. Katzenstein, and Luigi Graziano (New York: Praeger, 1978), 106.

23. Padioleau, *Quand la France s'enferre*, 124.

24. Ferry later claimed that his leadership position in the Fos project derived from a concern to balance the rivalries between the Usinor and Wendel-Sidélor. See Philippe Mioche, *Jacques Ferry et la sidérurgie française depuis la Seconde Guerre Mondiale* (Aix-en-Provence: Publications de l'Université de Provence, 1993), 214.

25. John H. McArthur and Bruce R. Scott, *Industrial Planning in France* (Boston: Harvard Graduate School of Business, 1969), 170.

26. Marger, *La situation et l'avenir de la sidérurgie*, 9.

27. See, for instance, Paul Laffont, "Les entreprises sidérurgiques depuis cinq ans," *Economie et Statistique*, no. 27 (October 1971). The figures in this paragraph come from Laffont.

28. Because of the joint consultation clause in the Convention, the CFDT had access to the individual contracts between the firms and the state. The 1966 figure comes from a copy of the Usinor-state contract filed in the CFDT-FGM archives: box "Sidérurgie CECA," file "Conventions particuliers."

29. Alain Lefour, "Wendel-Sidélor: Le prix de la mutation," *Entreprise*, no. 847 (December 3, 1971); Monique Thouvenin, "De Wendel-Sidélor à Sacilor-Sollac: 1968–1980. 13 Années de mutations difficiles en Lorraine," *Revue géopraphique de l'Est*, nos. 1–2 (January–June 1981).

30. The 1966 figure comes from a copies of the individual contracts between the state and de Wendel, Sidélor, and the SMS filed in the the CFDT-FGM archives: box "Sidérurgie CECA," file "Conventions particuliers." The 1970 figure is from Wendel-Sidélor, Rapport d'activité," 1970.

31. Commissariat Général du Plan, d'Équipement et de la Productivité, "Exécution de la Convention État-Sidérurgie," n.d. The Nord also had an easier time, since it only had to shed 1400 jobs.

32. In the late 1960s, Saint-Gobain-Pont-à-Mousson, holder of half the stock of Wendel-Sidélor, decided to withdraw entirely from steelmaking and focus its efforts on pipes and plastics. See Roger Martin, *Patron de droit divin* (Paris: Gallimard, 1984).

33. Sollac also kept close commercial ties to Dillingen Hütte in the Saarland. Owned by Pont-à-Mousson, DH also had a major stake in Sollac. Thus, semi-finished products were sent to Sollac for rolling.

34. Michel Freysennet, *La sidérurgie française, 1945–1979: L'Histoire d'une faillite, Les solutions qui s'affrontent* (Paris: Savelli, 1979), 132.

35. Padioleau, *Quand la France s'enferre*, 203.

36. Ibid., 196.

37. Anthony Cockerill, *The Steel Industry: International Comparisons of Industrial Structure and Performance*, University of Cambridge, Department of Applied Economics, Occasional Paper no. 42 (London: Cambridge University Press, 1974), 43–45.

38. To a lesser extent, the new skill base also reflected access to vocational education: young men tended to enter the firm with a technical certicate.

39. Freysennet, *La sidérurgie française, 1945–1979*, 148–54.

40. Jacques Broda, Serge Demailly, and Chantal Labruyère, "Crise de la sidérurgie et recomposition du procès de travail: La sous-traitance à la Solmer," *Sociologie du travail* 20, no. 4 (1978): 423–47.

41. Unpublished documents, Usinor-Sacilor, reprinted in CFDT, "Informations des administrateurs," no. 198 (November 28, 1989).

42. Geneviève Gourc and Jacques Gourc, "La restructuration de la sidérurgie dans les usines performantes: la Solmer à Fos-sur-Mer," *Critiques de l'Économie Politique*, nos. 15–16 (1981): 135–76.

43. Chambre Syndicale de la Sidérurgie Française, *La sidérurgie française*, various years.

44. Salaries in steel ranked ahead of those of shipyard workers and general mechanics but well behind those in autos, aeronautics, computers, electromechanics, and telecommunications. See *Le Guide du militant de la métallurgie*, no. 148 (March 1980).

45. See CFDT-FGM, "Rapport général, Activités et perspectives fédérales," 33rd Congress, April 9–11, 1965.

46. See Marcel Caille, "Le problème de la sidérurgie," *Reveil ouvrier*, April 1968.

47. "Compte rendu du conseil syndical, élargi aux sections non représentées au Conseil, du 19 juin 1968 à Hagondange," Syndicat CFDT–Sidérurgie lorraine, Hagondange, July 5, 1968, in *L'Homme du fer*, ed. Bonnet et al., 3:260.

48. "Causons salaires avec Jean Breteau-Secrétaire Générale de la Fédération," *La vie ouvrière-métaux*, February 12, 1969.

49. In an internal document remarkable for its comprehensiveness, the CFDT-FGM justified its position on nationalization. See CFDT-FGM, Branche Sidérurgie, "Quelques réflexions sur une position CFDT face á la nationalisation de la sidérurgie," January 1968, 28 pp., in CFDT-FGM archives, box "Documents pour Comité National Sidérurgie 1968."

50. Lévy, "Salaires. Syndicats. Grèves. L'opinion des employés et ouvriers de la sidérurgie lorraine," I.F.O.P., October 1969, in *L'Homme du fer*, ed. Bonnet et al., 3:286.

51. CGT-FTM, Bureau Fédéral, "Communiqué," September 1966.

52. "La CFDT et le Plan Professionnel," CFDT-FGM archives, Box "Sidérurgie Plan Professionel," file "Contacts avec," handwritten notes by national secretariat.

53. CFDT-FGM, "Préparation à une réunion paritaire nationale sidérurgie, 21 mai 1966," CFDT-FGM archives, Box "Sidérurgie Plan Professionnel," file "Réunions Nationales."

54. A follow-up study of the convention concluded that the training provisions of the social plan were not effective because redundant personnel were waiting for better times in steel and refused to consider exiting the industry. See Commissariat Général du Plan d'Equipement et de la Productivité, "Exécution de la Convention Générale Etat-Sidérurgie en 1969."

55. "Discussions d'action suite à l'inter-branche sidérurgie du 6 janvier 1967" (notes on CFDT-CGT meeting of the same day), CFDT-FGM archives, box "Sidérurgie Plan Professionnel," file "Démarches."

56. Ibid.

57. For an analysis of the iron industry, see Jean-Louis Masson, *L'Industrie lorraine du fer . . . ou comment tuer la poule aux oeufs d'or* (Paris: Cujas, 1977). Employment reduction by the early 1960s was necessary because the steel industry was moving away from the use of the poor-quality Lorraine ore and importing richer from overseas. Likewise, the mechanization of the mines

which had begun in the early 1950s increased productivity over 100 percent in the following decade.

58. Iron mines existed elsewhere in France, in particular near the integrated steel complex in Mondeville outside Caen in Normandy. Yet the Lorraine pits produced the vast majority of iron ore.

59. This promise created some tension between the CSSF and the UIMM. See Mioche, *Jacques Ferry*, 196.

60. See *Convention sociale de la Sidérurgie Lorraine* and *Convention sociale du Nord*. Not surprisingly, where steel was not such a public issue, such as in the specialty steel mills in the Creusot area in central France, similar accords took years to be signed.

61. The details of the negotiations between the unions and the Société des Aciéries et Tréfileries de Neuves Maisons are a fascinating example of attempts to free-ride on existing agreements. The company proposed the package as part of a three-year investment program in conjunction with its participation in the Convention État-Sidérurgie. In order to make the new facilities more profitable, management calculated that it needed to run at a higher capacity than had previously been the case given the historical levels of industrial conflict. Management proposed a three-year contract that would guarantee employment to all workers for the contract's duration. The company promised to increase training in general and to sweeten the training provisions in the *Convention sociale*. It promised to increase salaries more than the increases granted in the regional collective bargaining session. In exchange, the company sought an agreement on strike activity. For labor market disputes, no strikes would take place until all possibilities of dialogue were used. For national strike orders, the unions could only call workers out if a majority voted in favor. While the local CFDT and FO were favorable to the deal, the local CGT rejected it. Federation approval became a moot point when Assimilor nixed the proposals. This story was put together from interviews in 1984–1985 with many of the participants.

62. "UIMM/Sidérurgie Est et Nord," handwritten notes of collective bargaining session, March 10, 1977, CFDT-FGM archives, box "Conseils Nationaux 1974/77, file "Session du 11 au 15 janvier (1977)."

63. "Convention générale entre l'état et l'industrie sidérurgique," July 1966, 15.

64. The CGT did not immediately sign the agreement because the agreement did not specify the same salary level for less work. The six hours less work was compensated at only 66 percent of previous real salary levels. Therefore, monthly incomes were not guaranteed by the pact, and the CGT disagreed with any approach that lowered incomes. Still, after some initial posturing, the CGT signed the accord, so it could be part of the union delegation monitoring the results. See François Sellier, "L'Évolution des négociations collectives dans la sidérurgie et la métallurgie en France," *Droit social* 33, nos. 9–10 (September–October 1970), 434.

65. Ibid., 433.

66. The full implications of *mensualisation* for the average worker were quite large. Monthly pay changed the calculation of vacation time, sick leave, absences, seniority, promotion, retirement, etc. It also removed social control from the supervisor's "ad hockery" and placed it in the realm of collective bargaining.

Chapter 6. The Politics of Industrial Decline

1. See Commissariat Général du Plan, Groupe Sectoriel d'Analyse et de Prévision, *Mines de Fer, Sidérurgie, Première Transformation de l'Acier* (Paris: Documentation Française, 1976); International Iron and Steel Institute, *Causes of the Mid–1970's Recession in Steel Demand* (Brussels: ISSI, April 1980).

2. C. Engerand and M. Perricaudet, "Bilan et perspectives de la sidérurgie française à l'orée du VIIe Plan," *Annales des Mines*, June 1976, 10.

3. See Michel Freysennet, *La sidérurgie française, 1945–1979: L'Histoire d'une faillite, Les solutions qui s'affrontent* (Paris: Savelli, 1979), 159–61; Michel Turin, "Qui contrôlera Creusot Loire?" *Entreprise*, no. 995 (October 4, 1974), 76–79.

4. Sacilor, "Rapport d'activité," 1975; Usinor, "Rapport d'activité," 1975. The German figure is from C. Engerand and M. Perricaudet, "Bilan et perspectives de la sidérurgie française," pp. 9–16.

5. Klaus Stegemann, *Price Competition and Output Adjustment in the European Steel Market* (Tübingen: Mohr, 1977), 343.

6. *Fortune*, June 1978.

7. L. Boy and A. Pirovano, "Dirigisme privé et dirigisme public dans l'industrie sidérurgique," in *Sidérurgie et droit de crise*, ed. L. Boy et al. (Paris: PUF, 1983), 65.

8. Jean G. Padioleau, *Quand la France s'enferre: La politique sidérurgique de la France depuis 1945* (Paris: PUF, 1981), 186.

9. This is Freyssenet's estimate of the redundancy costs. See *La sidérurgie française*, 189.

10. Diana Green, *Managing Industrial Change? French Policies to Promote Industrial Adjustment* (London: HMSO, 1981), 68–69, n. 26.

11. See Geneviève Gourdet, "Une forme spécifique de financement de la sidérurgie," in *Sidérurgie et droit de crise*, ed. Boy et al.

12. The CICC reorganized once again in September 1977. It split off its steelworking activities, becoming a holding company for three new companies. The Société Châtillon-Neuves-Maison produced crude steel and rolled both sectionals and strip in Neuves-Maisons, Longwy, Anzin, Isbergues, and Blagny. The Société des Aciers Spéciaux de la Chiers had three small plants that made specialty steel. The Société des Tréfileries et Câbleries de Chiers-Gorcy produced cables and netting in eight different plants.

13. Jean Cantenot, "Restructurer la sidérurgie française," *Projet*, no. 135 (May 1979).

14. A number of other smaller mills were closed: a wire mill in Anzin, a re-rolling mill in Blagny, a strip mill in Longwy-Chiers, etc. See Jacques Malézieux, "Crise et restructuration de la sidérurgie française: Le groupe Usinor," *L'Espace Géographique*, no. 3 (July–September 1980).

15. *Le Monde*, February 26, 1980, gave the estimate of FF 10 billion. The Rosenwald report, released in 1984, argued that the real cost was probably 10 percent higher.

16. See Jean Baumier, *La fin des maîtres de forges* (Paris: Plon, 1981), 160–69; and John Zysman, *Governments, Markets, and Growth: Financial Systems and the Politics of Industrial Change* (Ithaca: Cornell University Press, 1983), 157–60.

17. Jean-Louis Masson, *Demain l'acier français* (Paris: Cujas, 1986), 56–67.

18. Between March and September 1981—that is, both before and after the election of the Socialists—the MoF injected MF 4,800 of FDES credits to Usinor and Sacilor. See Masson, *Demain l'acier français*, 62.

19. See, for instance, Arezki Dahmani, "La sidérurgie: Le poids de l'assistance permanente," in *L'Industrie en France, Centre de Recherche en Économie Industrielle*, ed. Bertrand Bellon and Jean-Marie Chevalier (Paris: Flammarion, 1983), 121–56.

20. The quote is from a speech by President François Mitterrand to a colloquium funded by the Ministry of Research and Industry in November 1982. See *Une politique industrielle pour la France: Actes des journées de travail des 15 et 16 novembre 1982* (Paris: Documentation Française, 1983), 424.

21. The three plants were at Dunes (near Dunkerque), Valenciennes (also in the Nord), and Marais (in Saint-Etienne in central France).

22. The Société des Aciers Spéciaux de la Chiers had plants in Vireux-Malhain and Haute-Rivières in the Ardennes as well as in Corbie in the Somme.

23. Usinor organized two plants from Creusot-Loire in Ondaine and Pamiers (both in south-central France) under the subsidiary of Campagnie Française de Forges et Fonderie. The company produced higher-quality steel and rolled them into forged parts. Usinor created the Société Métallurgique de St.-Chély-D'Apcher to manage the St.-Chély plant purchased from Creusot-Loire which was also in south-central France. It specialized in ferro-alloys and rolled magnetic sheet. Finally, Usinor created Usinor-Inox to run Creusot-Loire's Firminy plant and a factory in eastern France in Pont-de-Roide, formerly owned by Peugeot-Loire. Usinor-Inox specialized in stainless steels.

24. Ugine Aciers belonged to the chemical and aluminum conglomerate Pechiney-Ugine-Kulhman. PUK had decided to get out of steelmaking by the late 1970s. See "La restructuration des aciers spéciaux: Les données économi-

ques," *Actualités Industrielles Lorraines*, no. 172 (May 1980): 2–12. Previously owned by Renault, the SAFE had long been a producer in Lorraine. It was the first French steel firm to invest in continuous casting.

25. Pierre Judet, *L'Evolution des débouches de la sidérurgie française—Perspectives à moyen terme*, report to the Ministry of Industry, March 1982.

26. Five of the 9 billion francs promised to Usinor went to Dunkerque. The production of heavy plate would also be transferred from Longwy to Dunkerque. See "Le tour d'Europe des sidérurgies: La grisaille," *Actualité Industrielle, Économique et Sociale*, no. 3 (September–October 1982), 24–26.

27. This estimate comes from testimony of Raymond Lévy before the Commission Production et Echanges of the National Assembly, January 25, 1984

28. Per ton costs were slightly higher in Denain, but the plant there could be serviced by barge, thereby lowering total costs. See Robert Denoyelle, "Usinor et Sacilor, le Nord et la Lorraine," *La Voix du Nord*, May 20, 1982.

29. Symbolic of this convergence, a leader of the CGT-FTM, Serge Valli, addressed the CFDT-FGM steel conference in Pont-à-Mousson in October 1974.

30. "Déclaration du Conseil syndical C.F.D.T. de la Sidérurgie lorraine, Hagondange, December 3, 1974, in *L'Homme du fer: Mineurs de fer et ouvriers sidérurgistes lorrains, 1974–1985*, ed. Serge Bonnet, with Etienne Kagan and Michel Maigret (Nancy and Metz: Presses Universitaires de Nancy and Éditions Serpenoise, 1985), 4:39.

31. "La crise? Une tromperie. Elle est artificielle," *Le Pavé* (plant journal of CFDT enterprise section, Sacilor-Gandrange-Rombas), December 1975, in *L'Homme du fer*, ed. Bonnet et al., 4:70.

32. Jean-Louis Moynot attributes the underemphasis of labor market action and the focus on political action, and especially within the CGT, to the period after May 1968. As confederal secretary of the CGT at the time, he was in a particularly advantageous position to evaluate the evolution of French trade unionism. See "Le mouvement syndical en mouvement," *Dialectiques* no. 28 (Autumn 1979), 52.

33. The PCF organized a referendum on the nationalization of the steel industry in Lorraine in January 1977. Not unexpectedly, a massive majority voted in favor of nationalization (94 percent of the 97,492 ballots cast). The lopsided nature of the results was facilitated by the wording of the ballot: approval meant a "YES for the guarantee of employment, economic development with the nationalization of the steel industry and the iron mines"; disapproval was worded simply "NO for the the nationalization of the steel industry and the iron mines." See Bonnet et al., eds., *L'Homme du fer*, 4:133.

34. *Rouge*, April 20, 1977, in ibid., 4:144.

35. See the reporting in the regional press—*Républicain Lorrain*, *Est Républicain*, and *La Voix du Nord*—between March and September 1978.

36. See Claude Durand, *Chômage et violence: Longwy en lutte* (Paris: Galilée, 1981), 20–23.

37. The perceptions of militants are drawn from a series of interviews in Lorraine and the Nord in 1984–1985.

38. David Charrasse, with Randolphe Di Sabatino and Denis Pierron, *Lorraine Coeur d'Acier* (Paris: Maspero, 1981); Gérard Noiriel, with Benaceur Azzaoui, *Vivre et lutter à Longwy* (Paris: Maspero, 1980).

39. Jean-Pierre Huiban, "The Industrial Counterproposal as an Element of Trade Union Strategy," in *The French Workers' Movement: Economic Crisis and Political Change*, ed. Mark Kesselman, with Guy Groux (London: Allen & Unwin), 234.

40. The memorandum can be found in a special issue, "Face à la crise de la sidérurgie: Les solutions de la F.T.M.-C.G.T pour une négociation immédiate," *Le guide du militant de la métallurgie*, no. 136 (November 1978).

41. "FTM-CGT, "Face à la crise," 80.

42. The federation's propositions did not break with strategic orientation. An attack on the 1977 Plan Acier had been published in November of that year, and Jacques Chérèque, reputed to be the heir apparent to Edmond Maire as secretary general of the confederation, introduced the book. See Jean-Yves Rognant, Camille Romain, and François Rosso, *Où va la sidérurgie?* (Paris: Syros, 1977). Rosso was secretary of the Usinor-Thionville CFDT local.

43. CFDT-FGM, "Avenir de la sidérurgie: Propositions de la FGM-CFDT."

44. See also the elaboration of the CFDT's social vision by the FGM's secretary general Georges Granger in "Produire: Pour qui? Pourquoi?" *Dialectiques*, no. 28 (Autumn 1979).

45. See George Ross, "The Perils of Politics: French Unions and the Crisis of the 1970s," in Peter Lange, George Ross, and Maurizio Vanicelli, *Unions, Change and Crisis: French and Italian Union Strategy and the Political Economy, 1945–1980* (London: George Allen & Unwin, 1982), 58–60. See also Guy Groux and René Mouriaux, *La C.F.D.T.* (Paris: Economica, 1989).

46. The federation in Paris had active support from locals in other steel regions—Briey, Nancy, Sollac, Denain—in its dispute with Longwy and Dunkerque. Communications between Paris and the locals can be followed in CFDT-FGM archives, box "Sidérurgie 1978/79 Marche sur Paris, Longwy, Denain, etc."

47. While some infiltration undoubtedly had taken place, there also existed sincere differences both over the union reaction to restructuring and relations with the CGT. For the locals, these two tactical differences conflated to an oppositionalist stance. See an interview with Robert Giovanardi, head of the CFDT-Longwy local in *Rouge*, March 15–22, 1979. This paragraph also draws on the author's interview with Giovanardi in 1984.

48. George Ross ("The Perils of Politics," 60–62) attributes the shift back

to oppositional unionism to these two events, both of which assumed a national political importance and which came to serve the interests of the Communist Party. This may have been true at the confederation. However, our evidence seems to show that in the FTM propositional unionism never won out completely in the first place and that it was finally defeated only in the early 1980s.

49. CGT, Centre Confédéral d'Études Économiques, "Crise et solution pour la sidérurgie," Paris, November 1979. Jean-Louis Moynot, as confederal secretary, had editorial control over this series, which may explain the consistency of positions in the steel industry.

50. For a discussion of the evolution of propositional thinking in the CGT, see the work of Philippe Zarifian, an economist who was instrumental in writing the Memorandum and who was a close collaborator of Moynot. See "Restructurer or not?" *Dialectiques*, no. 28 (Autumn 1979). See also "Tactiques de lutte dans la sidérurgie," *La nouvelle critique*, no. 123 (April 1979). For his perspectives on industrial policy in the steel industry, see "La Politique industrielle dans la sidérurgie française de 1977 à 1983: Le jeu de l'echec," *Cahier du Centre de Recherche sur les Mutations des Sociétés Industrielles*, no. 9 (August–September 1984).

51. Assimilor, *Note d'Information*, no. 3 (May 23, 1967).

52. The CGT seems to have had little knowledge about the mechanics of engaging the FNE in early retirement schemes. The first such accord in France actually involved the CFDT in negotiations at the Saint-Nazaire shipyards in 1965.

53. François Ceyrac et al., *La C.F.D.T. en questions: La réponse d'Edmond Maire* (Paris: Gallimard, 1984); Hervé Hamon and Patrick Rotman, *La deuxième gauche: Histoire intellectuelle et politique de la CFDT* (Paris: Éditions Ramsay, 1982).

54. See CFDT-FGM archives, Boxes "CGPS Sidérurgie #1" and "CGPS Sidérurgie #2."

55. See Eugène Descamps, *Militer: Une vie pour un engagement collectif* (Paris: Fayard, 1971). See also René Carême, *Combats d'un militant* (Paris: Éditions Ouvrières, 1977).

56. *Les Echos*, November 11, 1986. This figure must be placed in comparative perspective. FF 100 billion is equivalent to $16.6 billion if we use an average exchange rate of FF6/$1. In order to compare to the United States, by population, we would multiply that figure by 4, although this would be a conservative number. (Comparing steel capacities, the number might be 5.) French subsidies between 1978 and 1986 alone would be equivalent to over $65 billion!

Chapter 7. Competitive Success

1. Anthony Daley, "Parties, Unions, and Political Mobilization in France," in *The Mitterrand Era: Policy Alternatives and Political Mobilization in France*,

ed. Daley (Basingstoke and New York: Macmillan and New York University Press, 1995).

2. Thomas Grunert, "Decision-Making Processes in the Crisis Policy of the EEC: Neocorporatist or Integrationist Tendencies?" in *The Politics of Steel: Western Europe and the Steel Industry in the Crisis Years (1974–1984)*, ed. Yves Mény and Vincent Wright (Berlin and New York: de Gruyter, 1987), 222–307.

3. Seeking an interlocutor at the European level, the Commission played a role in the formation of Eurofer.

4. Dennis Swann, *Competition and Industrial Policy in the European Community* (London and New York: Methuen, 1983).

5. *L'Expansion*, November 8–21, 1985.

6. Construction-grade constitutes the largest portion of all specialty steels and consequently the highest sales. Sacilor also owned smaller specialty steel firms. Ugine-Gueugnon produced stainless plate, Ugine Aciers stainless sectionals, and Imphy S.A. super alloys. These companies are not in the Nord or Lorraine. Together they employed over 9,000 workers in 1984.

7. Much of this material comes from a confidential plan (not released to the press) assembled by the staffs of Sacilor and Usinor that were combined into Unimétal. Unimétal, "Plan 1984," September 1984.

8. The productivity disparities between sites for individual products in 1984 were enormous. For instance, it cost FF 2,522 in Gandrange and FF 2,434 at SMN to produce a ton of reinforcing bar, while it cost FF 1,867 at Alpa and FF 1,801 at Montereau. The latter are minimills. The cost differences were similar in other products. For rails it cost FF 3,833 per ton in Rombas and 2,505 in Hayange. In merchant bars Rombas was high, at FF 2,898 per ton, and Valenciennes low, at FF 2,323.

9. Unimétal, "Plan Unimétal, Mise à Jour," internal document, July 1985, photocopy. The plant at Trith was only operating at 37 percent capacity use. See *Le Monde*, July 31, 1985.

10. Ascométal, "Plan Ascométal: Mise à jour," internal document, July 1985.

11. Chérèque had coordinated the Lorraine steel unions for the CFDT in the 1960s.

12. European Coal and Steel Community, *Investment in the Iron and Steel Industry*, 1991.

13. First, convertible bonds from the Steel Intervention Fund were added to the registered capital of each company. Second, the capital was reduced to zero. Third, the registered capital was increased by emission of stock available to be purchased by the previous owners. Fourth, the capital was reduced to roughly 5 percent of its purchase price. The intent here was to provide a legal mechanism to erase contracted debts. In practice only the state bought the new stock.

Ironically, the stock market operation dispossessed those who had speculated in steel stocks under the Socialists. See *Le Monde*, September 21, 1986. Believing the state would continue to pour money into the two companies, investors bid up the value of the stocks. Between 1983 and August 1986, the value of Usinor's stock rose tenfold. *Libération*, November 11, 1986.

14. *Le Monde*, March 22, 1991.

15. *Informatique*, October 10, 1988.

16. Christine Agache and Michel Sueur, "Restructurations et qualifications: Le cas de la sidérurgie," *Revue d'économie industrielle*, no. 31 (1985). Sollac reduced input costs by 20 percent between 1985 and 1988. See *Dynasteurs*, November 1988.

17. Unimétal's "innovation Plan" sought to install an ethic of permanent innovation which would allow the company to tap into savings of up to one quarter of current costs. See *Républicain Lorrain*, November 8, 1988.

18. See International Labour Organisation, Iron and Steel Committee, *Productivity Improvement and Its Effects on the Level of Employment and Working Conditions in the Iron and Steel Industry* (Geneva: ILO, 1986), 38.

19. Giusto Barisi, *Les stratégies patronales et syndicales en matière de relations sociales face au processus d'automatisation: Le case de la sidérurgie nationalisée française comparé au cas italien* (Paris: Centre de Recherche sur les Mutations des Sociétés Industrielles, October 1984), 100–01.

20. Christian Apothéloz, " 'Topomaintenance': Sollac Fos innove," *Liaisons sociales mensuel*, no. 40 (June 1989).

21. Philippe Zarifian, "Qualification collective et automatisation: Le cas de la sidérurgie," *Formation emploi*, no. 1 (January–March 1983): 41.

22. The IRSID dates from attempts in the 1930s to coordinate research. The Vichy organization committee for steel authorized a research institute in 1943. The formal organization was founded in 1946 by the CSSF. It was funded by a state decree taxing each ton of steel and pig iron produced. Initially, IRSID sought innovations in production processes. Later it would pursue product development.

23. *L'Usine Nouvelle*, December 8, 1988.

24. Géraldine de Bonnafos, "Filières professionnelles et acquisition des compétences dans la sidérurgie," *Formation emploi*, no. 12 (October–December 1985).

25. Barisi, *Les stratégies patronales*, 24.

26. Chambre Syndicale de la Sidérurgie Française, *La sidérurgie française*, 1986.

27. Guy Groux and Catherine Lévy, "Savoir, travail et production: La sidérurgie à Thionville," *Travail*, no. 14 (October 1987): 63.

28. Daniel Fabbra, "Compétences: Un challenge en acier pour Sollac Florange," *Education permanente*, no. 105 (December 1990): 51.

29. Régine Bercot, "Transformations techniques et division du travail" Le cas des opérateurs dans la sidérurgie lourde," *Formation emploi*, no. 11 (July–September 1985): 37–47.

30. Philippe Zarifian, "Du taylorisme au systemisme: Une nouvelle approche de la qualification dans l'industrie," *G.I.P. Mutations Industrielles*, no. 8, July 15, 1987.

31. *L'Est Républicain*, June 21, 1988.

32. Guy Groux and Cathérine Lévy, "Mobilisation collective et productivité économique: le cas des cercles de qualité dans la sidérurgie," *Revue française de socioloqie* 26, no. 1 (January–March 1985): 72.

33. Georges Steines, *Le fer dans la peau* (Metz: Éditions Serpenoise, 1977), 98.

34. Giusto Barisi, "Implications des salariés et enjeux de la 'participation' dans les processus de restructuration: L'exemple de trois entreprises sidérurgiques," *Travail et emploi*, no. 23 (March 1985): 68.

35. This material on Solmer relies on a set of interviews conducted by the author in July 1989 and on the detailed report written by Giusto Barisi. See *Les stratégies patronales et syndicales*, 261–88.

36. Not only did Social Affairs train nearly 400 middle managers between 1982 and 1984, it developed its own expertise in quality circles which it later marketed to other companies.

37. Groux and Lévy, "Mobilisation collective," 82–83.

38. Ibid., 84.

39. Barisi, *Les stratégies patronales*, 126–28.

40. *Syndicalisme*, October 20, 1988.

41. *Libération*, February 11, 1988.

42. Stéphane Lupieri, "Les circles vicieux de la qualité," *Dynasteurs*, December 1988.

43. Groux and Lévy, "Savoir, travail et production."

44. Catherine Lévy, "Crise des organisations syndicales et politiques patronales," *Travail* 4 (April 1984): 33.

45. *Syndicalisme*, October 20, 1988.

46. Barisi, "Implications des salariés," 74.

47. *Républicain Lorrain*, February 28, 1990.

48. For representative statements, see the special issue, "Unité d'action: Un outil essentiel pour des luttes efficaces," in *Le Guide du militant de la métallurgie*, no. 149 (May 1980). See also a statement by Bernard Lamirand, federal secretary in charge of steel unions, in *CGT-FTM, Liaison Industrie—Sidérurgie*, no. 225 November 1–15, 1988.

49. CFDT-FGM, Conseil Fédéral, "Rapport et résolutions générale: Au coeur de la vie des luttes des changements," *Bulletin du militant*, special no. 206 (November 1980), reports and resolutions for the 38th Congress of the FGM, February 5–8, 1981.

50. The regional union in Hagondange published a set of propositions to counter the restructuring plan of Unimétal. The propositions provided only a minor critique of the choices made by the new subsidiary of Sacilor and Usinor. See *Infos-Sidérurgie*, no. 4 supplement (October 1984).

51. This bitterness degenerated to the point that a coalition of CFDT and CGC members of the works council at Saulnes et Uckange approved management's decision to fire a CGT works council member for having devoted too much time to council work. See *Le métallurgiste lorrain*, October 1980.

52. For a representative example of the CGT's position, see the interview with Gérard Lopparelli, a CGT member of the board at Sacilor, "La France ne produit pas trop d'acier," *L'Humanité*, July 5, 1983.

53. Christian de Montlibert, *Crise économique et conflits sociaux* (Paris: L'Harmattan, 1989).

54. Henri Benoits, Pierre Cours-Salies, and Daniel Lisembard, "Longwy: Quatre Militants," *Collectif*, no. 1 (Spring 1987): 5–16.

Chapter 8. Converting Interests in Steel

1. John H. McArthur and Bruce R. Scott, *Industrial Planning in France* (Boston: Harvard Graduate School of Business, 1969), 418.

2. This information on the MoI's relations to the trade unions in the 1980s comes from the author's interview with a *conseiller technique* in the MoI in April 1985.

3. Presidential elections took place in 1965, 1969, 1974, 1981, 1988. Legislative elections were held in 1967, 1968, 1973, 1978, 1981, 1986, 1988. Municipal elections occurred in 1965, 1971, 1977, 1983.

4. *Est Républicain*, November 21, 1971, cited in *L'Homme du fer: Mineurs de fer et ouvriers sidérurgistes lorrains, 1960–1973*, ed. Serge Bonnet, with Etienne Kagan and Michel Maigret (Nancy and Metz: Presses Universitaires de Nancy and Editions Serpenoise, 1984), 3:339.

5. André Sainjon, "Le gouvernement négociera-t-il avec les syndicats?" *Le Monde*, October 4, 1978.

6. The state focus of French unions in the 1970s was therefore only partially misplaced. By exchanging mobilization for electoral hopes, the CGT and the CFDT weakened their potential pressure on state institutions. However, their emphasis on political change targeted the real agent of industrial change. For a different opinion, see George Ross, "The Perils of Politics: French Unions and the Crisis of the 1970s," in Peter Lange, George Ross, and Maurizio Vanicelli, *Unions, Change and Crisis: French and Italian Union Strategy and the Political Economy, 1945–1980* (London and Boston: Allen & Unwin, 1982), 13–93.

7. See *Convention Sociale de la Sidérurgie Lorrain*, June 30, 1967.

8. The forty-hour-week clause effectively reduced the impact on real salaries. Since the average work week in 1967 was forty-eight hours, the compensa-

tion was really equivalent to 90 percent of 5/6 salary, or 75 percent of previous real salaries.

9. For a fuller analysis of the 1967 Convention sociale, see Jean-Daniel Reynaud, "La convention sociale de la sidérurgie lorraine," *Droit social*, April 1969, 219–27.

10. Suzanne Berger, "Lame Ducks and National Champions: Industrial Policy in the Fifth Republic," in *The Fifth Republic at Twenty*, ed. William G. Andrews and Stanley Hoffmann (Albany: State University of New York Press, 1981), 292–310.

11. This was not an entirely new concept. A collective bargaining round in 1974 considered this procedure the year before, but the employers refused to concede it to the unions.

12. Under the original system devised in 1969, the employer would pay 60 percent of the lost time while an insurance fund paid the remainder. Depending on the convention, the FNE might pay half of the employers' contribution. Thus, the FNE assumed the large responsibility of keeping workers from being laid off.

13. Michel Freysennet, *La sidérurgie française, 1945–1979: L'Histoire d'une faillite, Les solutions qui s'affrontent* (Paris: Savelli, 1979), 183.

14. See CSSF, *La sidérurgie française en 1978*.

15. *Convention de Protection Sociale de la Sidérurgie de l'Est et du Nord*, June 3, 1977.

16. In 1972 the Confederations negotiated a national accord for minimum resources due to workers over sixty years of age. This effectively guaranteed early retirement at that age at 70 percent of the previous gross. See M. M. Mohamed Salah, "Aspects sociaux de la crise sidérurgique," in *Sidérurgie et droit de crise*, ed. L. Boy et al. (Paris: PUF, 1983), 151–52.

17. With the *mensualisation* reforms, compensation to previous salary levels took on new meanings. In all likelihood, the average real salary compensation for early retirement was not significantly lower in 1977 than in 1967.

18. The *Convention de protection sociale* was signed June 3, 1977. As early as two months later, employers and state officials were emitting signals—softly, given the upcoming elections—that necessary redundancies would be double those that were announced.

19. Workers who lost their jobs through redundancies qualified for supplemental unemployment benefits (*allocation suplémentaire d'attente*—ASA), which boosted unemployment compensation to 90 percent of the previous gross salary. These benefits originated in a national collective bargaining agreement in 1974. Although paid through the UNEDIC, the supplements were co-financed via the FNE. This system depended on the certification of motif économique. Otherwise, workers received the standard base benefits.

20. "La convention générale de protection social du 24 juillet 1979," *Actualités Industrielles Lorraines*, October 1979.

21. "Répartition par âge du personnel ourvrier et ETAM des Sociétés Sollac-Sacilor et Usinor fin 1978," CFDT-FGM archives, box "CGPS Sidérurgie #2," file "Sidérurgie, réunions avec les ministres et préfets."

22. The social fund of the European Coal and Steel Community partially financed both measures.

23. The CGT did not initially sign the CGPS but noted several positive items. Six months later it signed. Signature brought participation in the tripartite institutions designed to monitor compliance.

24. See Service des Etudes et de la Statistique, Ministère des Affaires Sociales et de l'Emploi, "Quinze ans de politique de l'emploi. Dossier statistique, 1974–1988," April 1990, 98–100.

25. See Jean Rosenwald, "Rapport." Presented to the Minister of Industry, January 1984.

26. Article 56 of the ECSC treaty provides for "reconversion" aids to displaced workers. The monies pass through national governments to avoid policy conflicts between governments and the ECSC.

27. *Républicain Lorrain*, January 26, 1984.

28. The agreement was not signed by the CGT, which saw in the 1984 CGPS an acceptance of retrenchment in the steel industry. The other four unions signed.

29. See Marie-Claire Villeval et al., *La reconversion de la main d'oeuvre: Bilan des problématiques*, Report to the Ministère du Travail, de l'Emploi et de la Formation Professionelle (Paris: Documentation Française, 1990).

30. The term conversion has belonged to the French political lexicon since at least the early 1950s (especially in the coal and steel areas) to describe what later would be called structural adjustment. It was used explicitly as the rationale for the social fund of the European Coal and Steel Community. The first conversion fund—involving subsidized loans to displaced business—was created in 1954. Conversion constituted the driving force behind the creation of the Fonds National de l'Emploi in 1963. See Jean-Pierre Clamadieu, "Les actions de conversion: Éléments de bilan," *Cahiers du Centre de Recherches Administratives, Politiques et Sociales* (Université de Lille II), no. 8 (June 1989).

31. Individual workers also had the right to take the two years of compensation up front but at only 65 percent of salary if they had a viable project in which to invest. This capitalization led to the creation of numerous small businesses with high failure rates in the depressed regions. Roughly half of eligible workers took their compensation up front.

32. The FO claimed afterward that a variety of refinements dealing with payroll deductions increased the compensation to 90 percent of previous take-home salary. See Michel Huc, "La nouvelle convention de protection sociale de la sidérurgie," *Droit social* 47, no. 11 (November 1984): 641.

33. Groupe Sacilor, *Faits et chiffres*, April 25, 1986.

34. See the follow-up report of the CPGS for Lorraine, Convention Générale de Protection Sociale de la Sidérurgie du 24 Juillet 1984, *Bilan Lorrain de trois années de conversion*, 2 vols., July 1987.

35. For an analysis of the Longwy conversion experience, see David Charrasse, "L'Aménagement de la transition professionnelle des sidérurgistes du bassin de Longwy (1984–1989), in *Mutations industrielles et reconversion des salariés*, ed. Marie-Claire Villeval (Paris: l'Harmattan, 1992), 276–84.

36. Jacques Chérèque, "L'État-partenaire," *Économie et Humanisme*, no. 304 (November–December 1988): 15.

37. See Jean-Pierre Deck, *La reconversion c'est possible* (Toulouse: Érès, 1986); Jean-Luc Outin and Rachel Silvera, "Les entreprises dans la reconversion: L'exemple du Creusot," *Économie et Humanisme*, no. 304 (November–December 1988).

38. *Les Echos*, May 20, 1987.

39. *Républicain Lorrain*, January 7, 1990.

40. Bruno Vennin, "Conversion industrielle: L'État et les autres," Saint-Etienne: CRESAL, 1988.

41. *Républicain lorrain*, February 28, 1990.

42. Marie-Claire Villeval, "Se former ou capitaliser? Dynamique conventionnelle et trajectoires de reconversion dans une entreprise sidérurgique," Groupe de recherche sur l'éducation et l'emploi, Université de Nancy II, cited in *Libération*, June 27, 1990.

43. Participants could not take their compensation in lump sum form.

44. For the 1969 collective bargaining agreement on employment security, see Jean-Maurice Verdier, "Sur le nouveau droit de licenciement," *Droit social*, February 1976, 1–18. The 1986 legislation on the conversion convention was extraordinarily detailed. It was required for all layoffs up to ten persons, over ten in a firm of under fifty employees, or in firms without a works council. Otherwise, it was recommended. Note that firms with over fifty workers tend to have works councils, and these can invoke the intervention of similar tools in the required social plans that must be negotiated in the event of layoffs.

45. See Jean-Yves Hocquet, "La reconversion des salariés," *Regards sur l'actualité* 134 (October 1987): 34–46.

46. Pierre Marioni, "Les dispositifs de conversion," *Dossiers statistiques du travail et de l'emploi* 43–44 (October 1988): 66–67. See also Marioni, "Les congés de conversion," *Dossiers statistiques du travail et de l'emploi* 34–35 (October 1987): 107–14.

47. Marie-Claire Villeval, "Crise de la sidérurgie et reconversion des salariés," *Les Annales de la recherche urbaine*, nos. 35–36 (November 1987): 21–30.

48. Martine Burdillat, "La Reconversion des salariés: Éléments exploratoires," *GIP Mutations Industrielles*, no. 5 (April 30, 1987).

49. CFDT, "Information des administrateurs du Groupe Usinor Sacilor," no. 189, September 26, 1989.

50. See "Modernisation négociée des entreprises," *Liaisons sociales*, series D, 120/88 (December 14, 1988).

51. See, for instance, Michel Rocard and the Clubs Convaincre, *Construire le progrès social* (Paris: Syros, 1990).

52. Usinor, *Rapport social 1982*.

53. Usinor, *Rapport social 1983*.

54. Usinor-Sacilor, *Exercise 1987*.

55. Usinor-Sacilor, *Exercise 1988*.

56. *Usine Nouvelle*, July 13, 1988; *Le Monde*, July 7, 1988; *La Voix du Nord*, July 7, 1988. Unpublished data from the CFDT suggest that the percentage of steel workers helped through reindustrialization efforts continued to fall throughout the 1980s. In 1988 only 17 percent of all the jobs created in this manner went to former steel workers. See CFDT, "Information des administrateurs," no. 157 (December 13, 1988).

57. See *Le Monde*, October 6, 1986.

58. Anthony Daley, "Socialist Employment Policy in France: 1981–1993," *Studies in Political Economy* 42 (Autumn 1993): 7–43.

59. *Le Monde*, July 12, 1988.

60. *Les Echos*, February 8, 1989.

61. *Républicain Lorrain*, September 29, 1989.

62. M. V. Ticheur, "Le Pôle européen de développement," *Économie Lorraine*, no. 46 (May 1986); Francis Cuillier, "Longwy. Une impulsion nouvelle," *Économie Lorraine*, no. 73 (November 1988). For comparative reflections on European industrial districts, see Charles F. Sabel, "Flexible Specialisation and the Re-emergence of Regional Economies," in *Reversing Industrial Decline: Industrial Structure and Policy in Britain and Her Competitors*, ed. Paul Hirst and Jonathan Zeitlin (Oxford and New York: Berg, 1989), 17–70.

63. *Républicain Lorrain*, March 8, 1990.

64. *Républicain Lorrain*, February 22, 1990.

65. *Le Monde*, February 8, 1992.

66. CFDT-FGM archives, boxes "CGPS Sidérurgie #1" and "CGPS Sidérurgie #2."

67. *Intervention*, August 1983, 47, quoted in Peter A. Hall, *Governing the Economy: The Politics of State Intervention in Britain and France* (New York: Oxford University Press, 1986), 220.

68. For union frustration with the process of consultation in the 1960s and early 1970s, see CFDT-FGM archives, boxes "Sidérurgie Plan Professionnel (Point de vue national 1965–1967)" and "Sidérurgie CECA."

69. See W. Rand Smith, "The Mitterrand Government and Industrial Crisis: Restructuring in the Steel and Automobile Industries," in *The Mitterrand Era:*

Policy Alternatives and Political Mobilization in France, ed. Anthony Daley (Basingstoke and New York: Macmillan and New York University Press, 1995).

70. Vincent Hoffmann-Martinot and Pierre Sadran, "The Local Implementation of France's National Strategy," in *The Politics of Steel: Western Europe and the Steel Industry in the Crisis Years (1974–1984)*, ed. Yves Mény and Vincent Wright (Berlin and New York: de Gruyter, 1987), 534–92.

71. Most of the new jobs created by reindustrialization measures have been in small firms. Given the historical weakness of organized labor in small plants, the new jobs were a mixed blessing for the unions.

72. Jean-Paul Chagnollaud, Etienne Criqui, Louis-Philippe Laprévote, and Guy Ménard, *Les militants syndicaux en lorraine* (Nancy: Presses Universitaire de Nancy, 1983), 20.

73. For a general account of this phenomenon, see Dominique Labbé and Maurice Croisat, *La fin des syndicats?* (Paris: l'Harmattan, 1992).

74. Unpublished data; see "Informations des Administrateurs," no. 182 (July 3, 1989).

75. *Sidérurgie-Infos*, no. 26 (June 1990).

76. Interviews with the author in June–July 1989.

77. For a discussion of industry politics and labor relations in an earlier period of the French steel industry, see Serge Bonnet and Roger Humbert, *La ligne rouge des hauts fourneaux* (Paris: Denoël, 1981).

Chapter 9. The Denationalization of Industrial Politics in France

1. I was struck in interviews in the Nord and Lorraine by the extent to which both investments and capacity reductions were criticized by persons from areas where they did not take place. The regional competition was strong.

2. Scott Lash and John Urry, *The End of Organized Capitalism* (Madison: University of Wisconsin Press, 1987).

3. As Peter Katzenstein suggested, the small states face a greater dependence on the international market and therefore are neither able to impose tariff protection nor desirous of remaining tied to a small consumer market. See Peter J. Katzenstein, *Small States in World Markets: Industrial Policy in Europe* (Ithaca: Cornell University Press, 1985).

4. Jean-Jacques Carré, Paul Dubois, and Edmond Malinvaud, *La croissance française: Un essai d'analyse économique causale de l'après-guerre* (Paris: Seuil, 1972), 149.

5. John Zysman, *Governments, Markets, and Growth: Financial Systems and the Politics of Industrial Change* (Ithaca: Cornell University Press, 1983).

6. André Gauron, *Histoire économique et sociale de la Cinquième République: Le temps des modernistes*, vol. 1 (Paris: La Découverte, 1983).

7. Henry Ehrmann, *Organized Business in France* (Princeton: Princeton University Press, 1958).

8. Between 1954 and 1975 agricultural employment dropped by 60 percent. See William James Adams, *Restructuring the French Economy: Government and the Rise of Market Competition Since World War II* (Washington, D.C.: Brookings, 1989), 21. Landholdings increased in size such that by the early 1970s concentration was the highest in the Community. (See Carré et al., *La croissance française*.) Productivity increased dramatically between 1950 and 1980. The French rural sector, however, did not wither away. It was transformed, economically strengthened, and given political legitimacy both in national politics and within the Community.

9. Suzanne Berger, "The Traditional Sector in France and Italy," in Suzanne Berger and Michael J. Piore, *Dualism and Discontinuity in Industrial Societies* (Cambridge: Cambridge University Press, 1980).

10. François Morin, *La Structure financière du capitalisme français* (Paris: Calmann-Lévy, 1974).

11. Claude Courlet and Bernard Pecqueur, "Les systèmes industriels localisés en France: Un nouveau modèle de développement," in *Les régions qui gagnent. Districts et réseaux: Les nouveaux paradigmes de la géographie économique*, ed. Georges Benko and Alain Lipietz (Paris: PUF, 1992), 81–102.

12. See Robert Boyer, "Wage Labor, Capital Accumulation, and the Crisis, 1968–82," in *The French Workers' Movement: Economic Crisis and Political Change*, ed. Mark Kesselman, with Guy Groux (London: Allen & Unwin, 1984), 17–38; Boyer, 1987: "The Current Economic Crisis: Its Dynamics and Its Implications for France," in *The Mitterrand Experiment: Continuity and Change in Modern France*, ed. George Ross, Stanley Hoffmann, and Sylvia Malzacher (New York: Oxford University Press, 1987).

13. Chris Howell and Anthony Daley, "Introduction: The Transformation of Political Exchange," *International Journal of Political Economy* 22, no. 4 (Winter 1992–93): 3–16.

14. Andrew Glyn, Alan Hughes, Alain Lipietz, and Ajit Singh, "The Rise and Fall of the Golden Age," in *The Golden Age of Capitalism*, ed. Stephen A. Marglin and Juliet B. Schor (Oxford: Clarendon Press, 1990).

15. Telesis, *Competing for Prosperity: Business Strategies and Industrial Policies in Modern France* (London: Policy Studies Institute, 1986).

16. Michel Crozier, "Western Europe," in *The Crisis of Democracy*, ed. Crozier, Samuel Huntington, and Joji Watanuki (New York: New York University Press, 1975).

17. Alain Peyrefitte, *Le mal français* (Paris: Plon, 1977); George Ross, "Adieu vieilles idées: The Middle Strata and Decline of Resistance-Liberation Left Discourse in France," in *Contemporary France*, ed. Jolyon Howorth and George Ross (London: Pinter, 1987); George Ross, "French Intellectuals from Sartre to Soft Ideology," in *Intellectuals and Politics: Social Theory in a Changing World*, ed. Charles C. Lemert (Newbury Park: Sage, 1991).

18. See Stephen S. Cohen and Peter A. Gourevitch, eds., *France in the Troubled World Economy* (London: Butterworth, 1982).

19. André Gauron, *Histoire économique et sociale de la Cinquième République: Années de rêves, années de crises (1970–1981)*, vol. 2 (Paris: La Découverte-Maspero, 1988).

20. Élie Cohen, *L'État brancardier: Politiques du déclin industriel (1974–1984)* (Paris: Calmann-Lévy, 1989).

21. David Cameron, "The Colors of a Rose: On the Ambiguous Record of French Socialism," Center for European Studies Working Paper, 1988; Michel Beaud, *La politique économique de la gauche*, vol 1: *Le mirage de la croissance* (Paris: Syros, 1983); vol. 2: *Le grand écart* (Paris: Syros, 1985).

22. For an exploration of the domestic constraint of floating exchange rates, see Michael Loriaux, *France After Hegemony: International Change and Financial Reform* (Ithaca: Cornell University Press, 1991).

23. Peter A. Hall, *Governing the Economy: The Politics of State Intervention in Britain and France* (New York: Oxford University Press, 1986).

24. Anthony Daley, "Socialist Employment Policy in France: 1981–1993," *Studies in Political Economy* 42 (Autumn 1993).

25. "L'Emploi, approches économique et sociale: Rapports du CNPF," *Liaisons sociales*, series D, no. 60/93 (June 23, 1993).

26. "L'évolution des revenus en France (1986–1989): 8e rapport du CERC," *Liaisons sociales*, series D, no. 97/90 (September 6, 1990).

27. Gérard Cornilleau, Pierre Marioni and Brigitte Roguet, 1990: "Quinze ans de politique de l'emploi," *Observations et diagnostics économiques*, no. 31 (April 1990), 91–120.

28. Bernard Irion, *L'Usine du futur: L'entreprise communicante et intégrée, report to the Commissariat Général du Plan* (Paris: Documentation Française, June 1990).

29. Martine Aubry, *Relations sociales et emploi*, report of the Social Relations and Employment Commission to the Commissariat Général au Plan (Paris: Documentation Française, June 1989).

30. Patrice Leuthold et al., *L'Entreprise et le syndicalisme* (Paris: CNRS-IRESCO, 1988).

31. The Socialists tried in 1982 to tighten controls, but they relented after considerable opposition from the business community and after the CFDT had signed a collective bargaining agreement on the issue. See Gérard Lachaise, "Les contrats à durée déterminé successifs," *Droit social*, nos. 9–10 (September–October 1989).

32. Vivien A. Schmidt, "Through 'More State' and 'Less State,' The French State Goes On: Patterns of Industrial Policy-Making in France 1981–1991," paper presented at the annual meeting of the American Political Science Association, Washington, D.C., August 1991; Daley, "Socialist Employment Policy."

33. Cameron, "The Colors of a Rose," 84.

34. Serge July, *Les Années Mitterrand: Histoires baroque d'une normalisation inachevée* (Paris: Grasset, 1986).

35. David Cameron, "Exchange Rate Politics in France: The Regime-Defining Choices of the Mitterrand Presidency," in Daley, ed., *The Mitterrand Era.*

36. Andrew Moravcsik, "Negotiating the Single Act: National Interests and Conventional Statecraft in the European Community," in *The New European Community: Decisionmaking and Institutional Change*, ed. Robert O. Keohane and Stanley Hoffmann (Boulder: Westview, 1991).

37. Wayne Sandholtz and John Zysman, "1992: Recasting the European Bargain," *World Politics* 42, no. 1 (October 1989), 95–128.

38. Organisation for Economic Cooperation and Development, *Employment Outlook* (Paris: OECD, July 1991).

39. Other unions include the Fédération de l'Education Nationale and the civil service unions which have sizable memberships by French standards. These density figures also include numerous nonaffiliated unions.

40. Ministère du Travail, de l'Emploi et de la Formation Professionnelle, *Dossiers statistiques du travail et de l'emploi*, no. 68 (April 1991).

41. Dominique Martin, ed., *Participation et changement social dans l'entreprise* (Paris: L'Harmattan, 1989); Renaud Sainsaulieu, ed., *L'Entreprise: Une affaire de société* (Paris: Presses de la Fondation Nationale des Sciences Politiques, 1990).

42. Gérard Adam, "L'Institutionalisation des syndicats: esquisse d'une problématique," *Droit social*, November 1983; Louis Dirn and Jean-Pierre Jaslin, "Crise ou renforcement du syndicalisme? Observations et diagnostics économiques," no. 12 (July 1985).

43. Mark Kesselman, "French Labor Confronts Technological Change: The Story of a Reform that Never Was," in *The Mitterrand Era*, ed. Daley.

44. Jacques Capdevielle, Hélène Y. Meynaud, and René Mouriaux, *Petits boulots et grand marché européen: Le travail démobilisé* (Paris: Presses de la Fondation Nationale des Sciences Politiques, 1990).

45. Fred Block, *Postindustrial Possibilities: A Critique of Economic Discourse* (Berkeley and Los Angeles: University of California Press, 1990).

46. Denis Bonvalot, "La politique et l'activité internationales de la CGT," in *Syndicalisme: Dimensions internationales*, ed. Guillaume Devin (Garenne-Colombes: Éditions Européenes Erasme, 1990).

47. *L'Humanité*, January 27, 1992.

48. Jean-Pierre Chauvel, Michel Pouliquen, and Pierre Chardon, "Maastricht: Le social sous traité," *La Vie ouvrière*, May 26, 1992, 18–25.

49. As evidence of this distrust, the FO established a Brussels office to gain independent information from the European Commission.

50. When Franz Steinkühler, head of IG Metall, suggested to *Le Figaro*

(February 11, 1992) that French unions had lost their influence because of their divisiveness and that they should unite into a single confederation, he was sharply rebuked by Blondel for attempting to impose a German model on France.

51. Annette Jobert, "La Négociation collective dans les entreprises multinationales en Europe," in *Syndicalisme: Dimensions internationales*, ed. Devin.

Bibliography

Archival Sources

Archives of the Department of Meurthe-et-Moselle (Nancy)
Fonds de Cabinet, Sous-série W
No. 950, 1020, 1104, 1131
Archives of the Department of Moselle (Metz)
Fonds de Cabinet (1944–1950), 151W
No. 149, 150, 151, 152, 190, 191
Archives of the Fédération Générale de la Métallurgie-CFDT
Boxes:
CGPS Sidérurgie #1; CGPS Sidérurgie #2; Conseils Nationaux 74/77; Documents pour Comité National Sidérurgie 1968; Sessions Nationales 1977, Conseils Nationaux; Sidérurgie 1e, 2e, 3e, 4e, 5e Plans; Sidérurgie 4e, 5e, 6e, 7e Plans; Sidérurgie 1978/79 Marche sur Paris, Longwy, Denain, etc.; Sidérurgie CECA; Sidérurgie Conflit Wendel-Sidélor Nov. 1971; Sidérurgie Plan Professionnel (Point de vue national 1965–1967); Sidérurgie Plan Professionnel Régions; Sidérurgie Réunions, Équipes, Animation et Conseil de Branche

Collective Agreements

Chambre Syndicale de la Sidérurgie Française and the Ministry of Finance, "Convention générale entre l'état et l'industrie sidérurgique," July 1966.

Convention de Protection Sociale de la Sidérurgie de l'Est et du Nord, June 3, 1977.

Convention Générale de Protection Sociale de la Sidérurgie de l'Est et du Nord du 24 juillet 1979, November 7, 1979.

Convention Générale de Protection Sociale de la Sidérurgie du 24 juillet 1984, August 1984.

Convention Sociale de la Sidérurgie Lorraine, June 30, 1967.

Convention Sociale du Nord, September 1, 1967.

Government Reports and Publications

Aubry, Martine. *Relations sociales et emploi.* Report of the Social Relations and Employment Commission to the Commissariat Général du Plan. Paris: Documentation Française, June 1989.

Auroux, Jean. *Les droit des travailleurs.* Report to the President and to the Prime Minister. Paris: Documentation française, September 1981.

Brunhes, Bernard, and Daniel Urbain. "Un service public de l'emploi pour les années 1990." Report to the Planning Commissioner. In Martine Aubry, ed., *Relations sociales et emploi*, 305–65 (Appendix).

Commissariat Général du Plan. "Résumé du Comité sidérurgique et mines de fer," 1973.

Commissariat Général du Plan, Commission de la Sidérurgie. "Investissements: Chatillon, Commentry, Neuves Maisons," August 1953.

Commissariat Général du Plan, Commission de la Sidérurgie. "Investissements: La Chiers—Usine de Longwy." August 1953.

Commissariat Général du Plan, Commission de la Sidérurgie. "Investissements: Knutange." August 1953.

Commissariat Général du Plan, Commission de la Sidérurgie. "Investissements: Lorraine-Escaut." August 1953.

Commissariat Général du Plan, Commission de la Sidérurgie. "Investissements: La Normandie." August 1953.

Commissariat Général du Plan, Commission de la Sidérurgie. "Investissements: Pompey." August 1953.

Commissariat Général du Plan, Commission de la Sidérurgie. "Investissements: La Providence—Usine de Rehon." August 1953.

Commissariat Général du Plan, Commission de la Sidérurgie. "Investissements: Sidélor." August 1953.

Commissariat Général du Plan, Commission de la Sidérurgie. "Investissements: U.C.P.M.I.." August 1953.

Commissariat Général du Plan, Commission de la Sidérurgie. "Investissements: de Wendel." August 1953.

Commissariat Général du Plan, Commission de la Sidérurgie. "Sollac: Programme d'Investissements." July 1, 1953.

Commissariat Général du Plan, d'Équipement et de la Productivité. "Exécution de la Convention État-Sidérurgie." n.d.

Commissariat Général du Plan, Groupe Sectoriel d'Analyse et de Prévision. *Mines de Fer, Sidérurgie, Première Transformation de l'Acier.* Paris: Documentation Française, 1976.

Cour des Comptes, *Rapport au Président de la République*, 1989.

Français, Ariel. Les rapports entre l'état et la sidérurgie au cours des dix dernières années." Commissariat Général du Plan, September 1974.

Irion, Bernard. *L'Usine du futur: L'entreprise communicante et intégrée.* Report to the Commissariat Général du Plan. Paris: Documentation Française, June 1990.

Judet, Pierre. *L'Evolution des débouches de la sidérurgie française—Perspectives à moyen terme.* Report to the Ministry of Industry, March 1982.

Lemaire, M. "Rapport d'information sur la situation dans les mines de fer et la sidérurgie et les perspectives de l'emploi dans les régions concernées." Assemblée Nationale, First Ordinary Session 1966/67, no. 2113, October 28, 1966.

Lévy, Raymond. Testimony before the *Commission Production et Echanges* of the National Assembly, January 25, 1984.

Marger, Pierre-Louis. "La situation et l'avenir de la sidérurgie." Report presented to the Economic and Social Council, July 3–4, 1979.

Rosenwald, Jean. "Rapport." Presented to the Minister of Industry, January 1984.

Séguin, Philippe. "Rapport à la Commission des Lois Constitutionelles," Assemblée Nationale, Second Ordinary Session 1977/78, No. 417, June 22, 1978.

Service des Etudes et de la Statistique, Ministère des Affaires Sociales et de l'Emploi. "Quinze ans de politique de l'emploi. Dossier statistique, 1974–1988, April 1990.

Villeval, Marie-Claire, et al. *La reconversion de la main d'oeuvre: Bilan des problématiques.* Report to the Ministère du Travail, de l'Emploi et de la Formation Professionelle. Paris: Documentation Française, 1990.

Other Reports and Publications

Convention Générale de Protection Sociale de la Sidérurgie du 24 Juillet 1984. *Bilan Lorrain de trois années de conversion*, 2 Vols. July 1987.

European Coal and Steel Community, High Authority. *General Report of the Activities of the Community*, various years.

International Labour Organisation, Iron and Steel Committee. *Productivity Improvement and Its Effects on the Level of Employment and Working Conditions in the Iron and Steel Industry.* Geneva: I.L.O., 1986.

———. *Recent Developments in the Iron and Steel Industry.* Geneva: I.L.O., 1986.

———. *Training in the Iron and Steel Industry.* Geneva: I.L.O., 1992.

Lutz, Burkart, and Alfred Willener. *Niveau de mécanisation et mode de rémuneration.* Luxembourg: European Coal and Steel Community, High Authority, May 1960.

Organization for Economic Cooperation and Development. *Economic Survey: France 1988/1989.* Paris: OECD, 1989.

———. *Economic Survey: France 1989/1990.* Paris: OECD, 1990.

———. *Economic Survey: France 1990/1991*. Paris: OECD, 1991.

———. *Economic Survey: France 1991/1992*. Paris: OECD, 1992.

———. *Employment Outlook*. Paris: OECD, July 1991.

United Nations. Economic Commission for Europe. *Evolution of the Specific Consumption of Steel*. New York: United Nations, 1984.

———. *Structural Changes in the Iron and Steel Industry*. New York: United Nations, 1979.

Periodicals

Actualité Industrielle, Économique et Sociale; *Actualités industrielles lorraines*; *Dynasteurs*; *Les Echos*; *Économie Lorraine*; *Est Républicain*; *L'Expansion*; *Financial Times*; *Fortune*; *L'Humanité*; *Informatique*; *Liaisons sociales*; *Libération*; *Lorraine-Magazine*; *Le Monde*; *Républicain Lorrain*; *Rouge*; *Usine Nouvelle*; *La Vie Française*; *La Voix du Nord*

Statistical Sources

Chambre Syndicale de la Sidérurgie Française. *La sidérurgie française*, various years.

European Coal and Steel Community. *Investment in the Community Coalmining and Iron and Steel Industries*, various years.

———. *Investment in the Iron and Steel Industry*, various years.

———. *Report on Competition Policy*, various years.

Eurostat. *Coût de la Main d'Oeuvre*, various years.

———. *Iron and Steel, 1952–1982*.

———. *Iron and Steel Statistical Yearbook*, various years.

Groupement de l'Industrie Sidérurgique. *Rapport annuel*, various years.

International Iron and Steel Yearbook. *Steel Statistical Yearbook*, various years.

International Labour Organisation. *Yearbook of Labour Statistics*, various years.

Ministre de l'Emploi. *Comptes de l'emploi: Données physico-financières, 1973–1980*. Paris: Documentation Française, 1982.

Organization for Economic Cooperation and Development. *The Iron and Steel Industry*, various years.

———. *Main Economic Indicators*, various years.

Steel Industry Publications

Annual Reports: Lorraine-Escaut; Sacilor; Sidélor; Usinor; Usinor-Sacilor; de Wendel; Wendel-Sidélor.

Ascométal. "Plan Ascométal: Mise à jour." Internal Document, July 1985.

Assimilor. *Note d'Information*, various issues.

Chambre Syndicale de la Sidérurgie Française. "La sidérurgie française et le problème du logement." Paris, n.d.

Chambre Syndicale de la Sidérurgie. "La verité sur la sidérurgie." Paris, September 1978.

Groupe Sacilor. *Faits et chiffres*, various issues.

Unimétal. "Plan 1984." Internal Document, September 1984.

Unimétal, "Plan Unimétal, Mise à Jour," Internal document, July 1985.

Union Bulletins, Newsletters, and other Publications

Confédération Française des Travailleurs Chrétiens (CFTC) / Confédération Française Democratique du Travail (CFDT):

CFDT Aujourd'hui

Syndicalisme

CFTC/CFDT—Fédération Générale de la Métallurgie:

"Avenir de la sidérurgie: Propositions de la FGM-CFDT," 1979.

Bulletin du militant; Congrès fédéraux, Proceedings, 1965–1985.

FGM-Info; "Informations des administrateurs du Groupe Usinor Sacilor."

Infos-Sidérurgie

Voix des métaux

CFTC/CFDT—Lorraine

L'Insurgé du crassier

La Lorraine sociale

Présence CFDT

Sidérurgie Lorraine

Conféderation Générale du Travail (CGT):

Le Peuple

Centre Confédéral d'Études Économiques. "Crise et solution pour la sidérurgie." Paris, November 1979.

"Face à la crise de la sidérurgie: Les solutions de la F.T.M.-C.G.T pour une négociation immédiate." *Le guide du militant de la métallurgie*, no. 136 (November 1978).

CGT—Fédération des Travailleurs de la Métallurgie:

Congrès fédéral, 1946–1984

Guide du métallurgiste

Le guide du métallurgiste, sidérurgie

Le Guide du militant de la métallurgie

Liaisons Industries-Sidérurgie

La vie ouvrière

La vie ouvrière-métaux

Voix des métallurgistes

CGT—Lorraine:

Bulletin de l'Union Départementale de Meurthe-et-Moselle

Informations lorraines

Le métallurgiste lorrain

Reveil ouvrier
Le travailleur Mosellan
CGT—Nord:
Le phare du sidérurgiste; *Le Travailleur du Nord*
Conféderation Générale du Travail—Force Ouvrière (FO):
La métallurgie syndicaliste; *La voix du métallurgiste*
FO—Lorraine:
La métallurgie syndicaliste; *Le sidérurgiste lorrain*,

Secondary Sources

Acs, János. "The Development of the Steel Industry From the Viewpoint of Innovation Theory." In Walter H. Goldberg, eds., *Ailing Steel: The Transoceanic Quarrel*. Hants, England: Gower, 1986, 283–306.

Adam, Gérard. "De la C.F.T.C. à la C.F.D.T." *Revue française de science politique* 15, no. 1 (February 1965): 87–103.

———. "L'institutionnalisation des syndicats, esquisse d'une problématique." *Droit Social* 46, no. 11 (November 1983): 597–600.

———. *Le pouvoir syndical*. Paris: Dunod, 1983.

———. "Les syndicats: Un pouvoir excessif?" *Droit Social* 47, no. 1 (January 1984): 2–4.

Adam, Gérard, Frédéric Bon, Jean Capdevielle, and René Mouriaux. *L'ouvrier français en 1970*. Paris: A. Colin, 1971.

Adam, Gérard, and Jean-Daniel Reynaud. *Conflits du travail et changement social*. Paris: PUF, 1978.

Adam, Gérard, Jean-Daniel Reynaud, and Jean-Maurice Verdier. *La négociation collective en France*. Paris: Éditions Ouvrières, 1972.

Adam, Gérard, and Michel Lucas. "Les institutions de représentation du personnel en France: Bilan et perspectives." *Droit social* 39, no. 3 (March 1979): 79–91.

Adams, William James. *Restructuring the French Economy: Government and the Rise of Market Competition Since World War II*. Washington, D.C.: Brookings, 1989.

Adams, William James, and Christian Stoffaës, eds. *French Industrial Policy*. Washington, D.C.: Brookings, 1986.

Agache, Christine, and Michel Sueur. "Restructurations et qualifications: Le cas de la sidérurgie." *Revue d'économie industrielle*, no. 31 (1985).

Aglietta, Michel, and Anton Brender. *Les métamorphoses de la société salariale: La France en projet*. Paris: Almann-Lévy, 1984.

Amadieu, Jean-François. "Les tendances au syndicalisme d'entreprise en France: Quelques hypothèses." *Droit social* 49, no. 6 (June 1986): 495–500.

———. "Vers un syndicalisme d'entreprise: D'une définition de l'entreprise à celle du syndicalisme." *Sociologie du travail* 28, no. 3 (1986): 237–50.

Amadieu, Jean-François, and Nicole Mercier. "Le débat sur la 'flexibilité.' " *Sociologie du travail* 28, no. 2 (1986): 193–201.

Ambler, John S. "Educational Pluralism in the French Fifth Republic." In George Ross and James F. Hollifield, eds., *Searching for the New France*. New York and London: Routledge, 1991.

Amicabile, Alain. "Sidérurgie." *Économie et politique*, no. 65 (September 1982): 37–39.

Andrews, William G., and Stanley Hoffmann, eds. *The Fifth Republic at Twenty*. Albany: State University of New York Press, 1981.

Armingeon, Klaus, et al. *Les Syndicats européens et la crise*. Grenoble: Presses Universitaires de Grenoble, 1981.

Atkinson, Michael M., and William D. Coleman. "Corporatism and Industrial Policy." In Alan Cawson, ed., *Organized Interests and the State: Studies in Meso-Corporatism*. London and Beverly Hills: Sage, 1985, 22–44.

Aylen, Jonathan. "Plant Size and Efficiency in the Steel Industry: An International Comparison." *National Institute Economic Review*, no. 100 (May 1982): 65–76.

Baglioni, Guido, and Colin Crouch, eds. *European Industrial Relations: The Challenge of Flexibility*. London: Sage, 1990.

Barisi, Giusto. *Les stratégies patronales et syndicales en matière de relations sociales face au processus d'automatisation: Le case de la sidérurgie nationalisée française comparé au cas italien*. Paris: Centre de Recherche sur les Mutations des Sociétés Industrielles, October 1984.

———. "Implications des salariés et enjeux de la 'participation' dans les processus de restructuration: L'exemple de trois entreprises sidérurgiques." *Travail et emploi*, no. 23 (March 1985).

Barnett, Donald, and Louise Schorsch. *Steel: Upheaval in a Basic Industry*. Cambridge: Ballinger, 1983.

Baumgartner, Frank. *Conflict and Rhetoric in French Policy Making*. Pittsburgh: University of Pittsburgh Press, 1989.

Baumier, Jean. *La fin des maîtres de forges*. Paris: Plon, 1981.

Beaud, Michel. *La politique économique de la gauche: Le mirage de la croissance*, Vol. 1. Paris: Syros, 1983.

———. *La politique économique de la gauche: Le grand écart*, vol. 2. Paris: Syros, 1985.

Belleville, Pierre. "La sidérurgie lorraine et son prolétariat." *Les Temps modernes*, no. 191 (April 1962): 1492–1535.

———. *Laminage continu: Crise d'une région, échec d'une régime*. Paris: Julliard, 1968.

Bellon, Bertrand, and Jean-Marie Chevalier, eds. *L'Industrie en France*. Paris: Flammarion, 1983.

Benoits, Henri, Pierre Cours-Salies, and Daniel Lisembard. "Longwy: Quartre Militants," *Collectif*, No 1 (Spring 1987): 5–16.

Beranger, Thierry. "1959–1974: quinze ans d'assurance chômage," *Droit social*, (September–October 1975): 135–41.

Bercot, Régine. "Transformations techniques et division du travail" Le cas des opérateurs dans la sidérurgie lourde." *Formation emploi*, no. 11 (July–September 1985).

Berger, Suzanne. *Peasants Against Politics: Rural Organization in Brittany, 1911–1967*. Cambridge: Harvard University Press, 1972.

———. "The Traditional Sector in France and Italy." In Berger and Michael J. Piore, *Dualism and Discontinuity in Industrial Societies*. Cambridge: Cambridge University Press, 1980.

———. Lame Ducks and National Champions: Industrial Policy in the Fifth Republic." In Andrews and Hoffmann, eds., *The Fifth Republic at Twenty*, 292–310.

Bergounioux, Alain. *Force ouvrière*. Paris: Seuil, 1975.

Bergounioux, Alain, et al. *La parole syndicale: Étude du vocabulaire confédéral des centrale ouvrières françaises, 1971–1976*. Paris: PUF, 1982.

Betting, Jean-Baptiste. *Jean-Baptiste, Battre le fer*. Paris: Éditions ouvrières, 1982.

Biard, Roger. *La sidérurgie française*. Paris: Editions Sociales, 1958.

Birnbaum, Pierre. *La Logique de l'état*. Paris: Fayard, 1982.

———. "The State versus Corporatism. *Politics and Society* 11, no. 4 (1982): 477–501.

Blanc, Jacques, and Chantal Brulé. *Les nationalisations françaises en 1982*. Notes et Étude Documentaires, nos. 4721–22. Paris: Documentation française, 1983.

Bloch-Lainé, François. *Pour une réforme de l'entreprise*. Paris: Seuil, 1963.

Block, Fred. *Postindustrial Possibilities: A Critique of Economic Discourse*. Berkeley and Los Angeles: University of California Press, 1990.

de Bodman, Eric, and Bertrand Richard. *Changer les relations sociales: La politique de Jacques Delors*. Paris: Éditions d'Organisation, 1976.

de Bonnafos, Géraldine."Filières professionnelles et acquisition des compétences dans la sidérurgie." *Formation emploi*, no. 12 (October–December 1985).

Bonnefont, Jean-Claude. *La Lorraine*. Paris: PUF, 1984.

Bonnet, Serge. "Les ouvriers migrants quotidiens des usines sidérurgiques de l'agglomération de Longwy, 1962–1963." Diss. Faculté des Lettres de Paris, 1965.

———. "Political Alignments and Religious Attitudes within the Italian Immigration to the Metallurgical Districts of Lorraine." *Journal of Modern History* 2, no. 2 (Winter 1968): 123–55.

———. *Sociologie politique et religieuse de la Lorraine*. Cahier de la Fondation Nationale des Sciences Politique, no. 181. Paris: A. Colin, 1972.

Bonnet, Serge, and Roger Humbert. *La ligne rouge des hauts fourneaux*. Paris: Denoël, 1981.

Bonnet, Serge, with Etienne Kagan and Michel Maigret, eds. *L'Homme du fer: Mineurs de fer et ouvriers sidérurgistes lorrains, 1930–1959*, Vol. 2. Nancy: Centre Lorrain d'Études Sociologiques, 1977.

Bonnet, Serge, with Etienne Kagan and Michel Maigret, eds. *L'Homme du fer: Mineurs de fer et ouvriers sidérurgistes lorrains 1960–1973*, Vol. 3. Nancy and Metz: Presses Universitaires de Nancy and Éditions Serpenoise, 1984.

Bonnet, Serge, with Etienne Kagan and Michel Maigret, eds. *L'Homme du fer: Mineurs de fer et ouvriers sidérurgistes lorrains 1974–1985*, Vol. 4. Nancy and Metz: Presses Universitaires de Nancy and Éditions Serpenoise, 1985.

Bonnet, Serge, Charles Santini, and Hubert Barthelemy. "Des prolétaires catholiques aux notables communistes." *Esprit*, no. 348 (April 1966): 826–37.

Bonvalot, Denis. "La politique et l'activité internationales de la CGT." In Guillaume Devin, ed., *Syndicalisme: Dimensions internationales*. Garenne-Colombes: Éditions Européenes Erasme, 1990.

Bouffartigue, Paul, and Chantal Labruyère. "Les cercles de qualité: La portée d'une implantation réussie—La Solmer à Fos-sur-Mer," *Travail et emploi* 20 (June 1984): 45–56.

Boyer, Robert. "Wage Labor, Capital Accumulation, and the Crisis, 1968–82." In Kesselman with Groux, eds., *The French Workers' Movement*, 17–38.

———. "The Current Economic Crisis: Its Dynamics and its Implications for France." In George Ross, Stanley Hoffmann, and Sylvia Malzacher, eds. *The Mitterrand Experiment: Continuity and Change in Modern France*. New York: Oxford University Press, 1987.

———. *La Théorie de la régulation: Une analyse critique*. Paris: La Découverte, 1987.

Boy, L. et al., eds. *Sidérurgie et droit de crise*. Paris: PUF, 1983,

Boy, L., and A. Pirovano. "Dirigisme privé et dirigisme public dans l'industrie sidérurgique." In Boy et al., eds., *Sidérurgie et droit de crise*, 15–73.

Branciard, Michel. *Syndicats et partis: Autonomie ou dépendance, 1948–1981*, vol. 2. Paris: Syros, 1982.

Brenner, Robert, and Mark Glick. "The Regulation Approach: Theory and History." *New Left Review* 188 (July–August 1991): 45–119.

Bridgford, Jeff. "The Integration of Trade Union Confederations into the Social and Political System." In Cerny, ed. *Social Movements*.

Broda, Jacques, Serge Demailly, and Chantal Labruyère. "Crise de la sidérurgie et recomposition du procès de travail: La sous-traitance à la Solmer." *Sociologie du travail* 20, no. 4 (1978): 423–47.

Bron, Jean. *Histoire ouvrière: La CFDT dan l'Isère*. Grenoble: Presses Universitaires de Grenoble, 1984.

Browaeys, Xavier, and Paul Chatelain. *La France du travail*. Paris: PUF, 1984.

Bunel, Jean, and Jean Saglio. "La faiblesse de la négociation collective et le pouvoir patronal." *Sociologie du travail* 19, no. 4 (October–December 1977): 387–401.

Burdillat, Martine. "La Reconversion des salariés: Éléments exploratoires." G.I.P. Mutations Industrielles, no. 5 (April 30, 1987).

Caire, Guy. *Les syndicats ouvriers*. Paris: PUF, 1971.

———. *La grève ouvrière*. Paris: Éditions ouvrières, 1978.

Camdessus, Christine. *Les interventions financières des états en faveur de la sidérurgie dans la C.E.C.A*. Paris: PUF, 1987.

Cameron, David R. "The Colors of a Rose: On the Ambiguous Record of French Socialism." *Center for European Studies Working Paper*, no. 12, 1988.

———. "Exchange Rate Politics in France: The Regime-Defining Choices of the Mitterrand Presidency." In Daley, ed., *The Mitterrand Era*.

Cantenot, Jean. "Restructurer la sidérurgie française." *Projet*, no. 135 (May 1979): 632–38.

Capdevielle, Jacques, Elisabeth Dupoirier, and Guy Lorant. *La grève du Joint français: Les incidences politiques d'un conflit social*. Paris: FNSP / A. Colin, 1975.

Capdevielle, Jacques, and René Mouriaux. *Les syndicats ouvriers en France*. Paris: A. Colin, 1970.

Capdevielle, Jacques, and René Mouriaux, and Hélène Y. Meynaud. *Petits boulots et grand marché européen: Le travail démobilisé*. Paris: Presses de la Fondation Nationale des Sciences Politiques, 1990.

Carême, René. *Combats d'un militant*. Paris: Éditions Ouvrières, 1977.

Caron, François. *An Economic History of Modern France*. Trans. Barbara Bray. New York: Columbia University Press, 1979.

Carré, Jean-Jacques, Paul Dubois, and Edmond Malinvaud. *La croissance française: Un essai d'analyse économique causale de l'après-guerre*. Paris: Seuil, 1972.

Cartier, Raymond. "La Lorraine, Texas français," *Paris-Match*, June 26, 1961.

de Casanova, E. Arrighi. "Les quasi-contrats du Plan." *Droit Social* 28, no. 6 (June 1965): 341–55.

Catherine, Bernard. "La Chambre Syndicale de la Sidérurgie Française et la Convention État-Sidérurgie ou l'action d'un syndicat professionnel aujourd'hui." Unpublished paper. Institut d'Études Politiques, Grenoble, n.d.

Cattiaux, Guy. *Denain: Des hommes d'acier, Une région à sauver*. Denain: Guy Cattiaux, 1980.

Cawson, Alan, ed. *Organized Interests and the State: Studies in Meso-Corporatism*. London and Beverly Hills: Sage, 1985.

Cerny, Philip G., ed. *Social Movements and Protest in France*. New York: St. Martin's, 1982.

Ceyrac, François et al. *La C.F.D.T. en question. La réponse d'Edmond Maire.* Paris: Gallimard, 1984.

Chagnollaud, Jean-Paul, Etienne Criqui, Louis-Philippe Laprévote, and Guy Ménard. *Les militants syndicaux en Lorraine.* Nancy: Presses Universitaires de Nancy, 1983.

Chardonnet, Jean. *La sidérurgie française: Progrès ou décadence?* Cahier de la Foundation Nationale des Sciences Politiques, no. 63. Paris: A. Colin, 1954.

Charrasse, David with Randolphe Di Sabatino and Denis Pierron. *Lorraine Coeur d'Acier.* Paris: Maspero, 1981.

Charasse, David, and Gérard Noiriel. "Marcel Donati: Ouvrier Lamineur" (interview). *Travail* 4 (April 1984): 25–27.

Chaskiel, Patrick, and Marie-Claire Villeval. *Pompey: Crise, Fermeture, Reconversion.* Nancy: Presses Universitaires de Nancy, 1988.

Chaskiel, Patrick, Hervé Lhotel, and Marie-Claire Villeval. "Négotiation et transformations du rapport salarial dans la crise d'une entreprise." In Robert Salais and Laurent Thévenot, eds., *Le Travail: Marchés, Règles, Conventions.* Paris: Economica, 1986.

Chatillon, Stéphane. *Pourquoi la division syndicale?* Paris: Économica, 1978.

Chérèque, Jacques. "L'État-partenaire." *Économie et Humanisme* 304 (November–December 1988).

Chetcuti, Claude. "Réflexions sur l'inspection du travail." *Droit Social* 39, no. 2 (February 1976), 19–37.

Clamadieu, Jean-Pierre. "Les actions de conversion: Éléments de bilan." *Cahiers du Centre de Recherches Administratives, Politiques et Sociales* (Université de Lille II), no. 8 (June 1989).

de Closets, François. *Toujours plus!* Rev. ed. Paris: Grasset, 1984.

Cockerill, Anthony. *The Steel Industry: International Comparisons of Industrial Structure and Performance.* University of Cambridge, Department of Applied Economics, Occasional Paper no. 42. London: Cambridge University Press, 1974.

Cohen, Élie. *L'État brancardier: Politiques du déclin industriel (1974–1984).* Paris: Calmann-Lévy, 1989.

Cohen, Stephen S. *Modern Capitalist Planning: The French Model*, 2d ed. Berkeley and Los Angeles: University of California Press, 1977.

Cohen, Stephen S., and Peter A. Gourevitch, eds. *France in a Troubled World Economy.* London: Butterworth, 1982.

Collectif. *Lip: affaire non classée.* Paris: Syros, 1976.

Coriat, Benjamin. *L'Atelier et le chronomètre: Essai sur le Taylorisme, le Fordisme et la production de masse*, 2d ed. Paris: C. Bourgois, 1982.

Cornilleau, Gérard, Pierre Marioni, and Brigitte Roguet. "Quinze ans de politique de l'emploi." *Observations et diagnostics économiques* 31 (April 1990): 91–120.

Courlet, Claude, and Bernard Pecqueur. "Les systèmes industriels localisés en France: Un nouveau modèle de développement." In Georges Benko and Alain Lipietz, eds., *Les régions qui gagnent. Districts et réseaux: Les nouveaux paradigmes de la géographie économique*. Paris: PUF, 1992, 81–102.

Cousin, Alexis. "Nationaliser la sidérurgie. Les licenciements chez Wendel-Sidélor. La Convention État-Sidérurgie." *Économie et politique*, December 1971, 129–35.

Couturier, Gérard. "Le plan social: Aspects juridiques." *Droit social* 48, nos. 9–10 (September–October 1985): 643–50.

Cox, Andrew, and Jack Hayward. "The Inapplicability of the Corporatist Model in Britain and France: The Case of Labor." *International Political Science Review* 4, no. 2 (1983): 217–40.

Cox, Robert W. "Pour une étude prospective des relations de production." *Sociologie du travail* 19, no. 2 (April–June 1977): 113–37.

Crouch, Colin, and Alessandro Pizzorno, eds. *The Resurgence of Class Conflict in Western Europe Since 1968*, 2 Vols. New York: Holmes and Meier, 1978.

Crozier, Michel. *The Bureaucratic Phenomenon*. Chicago: University of Chicago Press, 1964.

———. *La société bloquée*. Paris: Seuil, 1970.

———. "Western Europe." In Crozier, Samuel P. Huntington, and Joji Watanuki, *The Crisis of Democracy*. New York: New York University Press, 1975.

Dahmani, Arezki. "La sidérurgie: Le poids de l'assistance permanente." In Bertrand Bellon and Jean-Marie Chevalier, eds. *L'Industrie en France*. Centre de Recherche en Économie Industrielle. Paris: Flammarion, 1983, 121–56.

Dahl, Robert. *Democracy and Its Critics*. New Haven: Yale University Press, 1989.

Daley, Anthony. "Remembrance of Things Past: The Union-Party Linkage in France." *International Journal of Political Economy* 23, no. 4 (Winter 1992–93): 53–71.

———. "Socialist Employment Policy in France: 1981–1993." *Studies in Political Economy* 42 (Autumn 1993): 7–43.

———. "Parties, Unions, and Political Mobilization in France." In Daley, ed. *The Mitterrand Era*.

Daley, Anthony, ed. *The Mitterrand Era: Policy Alternatives and Political Mobilization in France*. Basingstoke and New York: Macmillan and NYU Press, 1995.

Dauba, Michel. "Sidérurgie: Ne pas ceder devant la C.E.C.A." *Économie et politique*, no. 67 (November 1982): 9–10.

Dauba, Michel, and Daniel Vergnaud. "Investir et produire en France pour acheter et vendre français." *Économie et politique*, nos. 63–64 (July–August 1982): 18–20.

De Angelis, Richard A. *Blue-Collar Workers and Politics: A French Paradox.* London: Croom Helm, 1982.

Deck, Jean-Pierre. *La reconversion c'est possible.* Toulouse: Érès, 1986.

Delamotte, Ives. "Recent Collective Bargaining Trends in France." *International Labour Review* 103, no. 4 (April 1971): 351–77.

Denoyelle, Robert. *Les feux éteints.* Denain: Guy Cattiaux, 1988.

Descamps, Eugène. *Militer: Une vie pour un engagement collectif.* Paris: Fayard, 1971.

Despax, Michel. "La place de la convention d'entreprise dans le système conventionnel." *Droit social* 51, no. 1 (January 1988): 8–16.

Diebold, William Jr. *The Schuman Plan: A Study in Economic Cooperation.* New York: Praeger, 1959.

Dion, Michel. *Les catholiques et le pouvoir: La crise du "consensus" (enquête en Mayenne et en Lorraine).* Paris: Éditions sociales, 1980.

Dirn, Louis, and Jean-Pierre Jaslin. "Crise ou renforcement du syndicalisme? *Observation et diagnostics économiques* 12 (July 1985).

Dofny, Jacques, Claude Durand, Jean-Daniel Reynaud, and Alain Touraine. *Les ouvriers et le progrès. Étude de cas: Un nouveau laminoir.* Paris: Armand Colin, 1966.

Dourille, Edmond. *La sidérurgie dans le monde depuis 1952.* Paris: Documentation Française, 1981.

Dubois, Pierre. *Mort de l'état-patron.* Paris: Éditions Ouvrières, 1974.

Dunlop, John T. *Industrial Relations Systems.* New York: Henry Holt, 1958.

Duprilot, Jean-Pierre, and Paul Fieschi-Vivet. *Les comités d'entreprise (Droit et pratique).* Paris: PUF-Que sais-je?, 1982.

Dupuy, Fernand. *Etre maire communiste.* Paris: Calmann-Lévy, 1975.

Durand, Claude. "L'Évolution du travail dans les laminoirs." *Revue française du travail* 13, no. 1 (January–March 1959): 3–18.

———. "Les ouvrier et le progrès technique: Mont-Saint-Martin vingt ans après." *Sociologie du travail* 22, no. 1 (January–March 1980): 4–21.

———. *Chômage et violence: Longwy en lutte.* Paris: Galilée, 1981.

———. "La violence à Longwy." *Sociologie du travail* 23, no. 2 (April–June 1981): 218–29.

———. "Les syndicats et la politique industrielle." *Sociologie du travail* 28, no. 3 (1986): 304–15.

Durand, Claude, and Ginette Lemaitre. "Usinor: Le Plan Acier, 1982–86." In Cécilia Cassussus et al., eds., *Stratégies industrielles et politiques d'emploi.* Paris: CNRS, 1984.

Durand, Michelle. "De l'analyse causale à l'analyse structurelle, la négociabilité des conflits." *Sociologie du travail* 19, no. 4 (October–December 1977): 402–21.

———. "La grève: conflit structurel, système de relations industrielles ou facteur de changement social. *Sociologie du travail* 21, no. 3 (July–September 1979): 274–97.

Durand, Paul. "Fin d'une controverse: Les effets de la grève sur le contrat de travail (L'article 4 de la loi du 11 février 1950)." *Sociologie du travail* 13, no. 3 (March 1950): 118–26.

———. "La loi du 11 février 1950 sur les conventions collectives du travail." *Droit social* 13, no. 3 (March 1950): 93–101.

———. "La loi du 11 février 1950 sur les conventions collectives du travail." *Droit social* 13, no. 4 (April 1950): 155–63.

———. "La loi du 11 février 1950 sur les conventions collectives du travail." *Droit social* 13, no. 5 (May 1950): 186–93.

Durrieu, Yves. *L'Impossible régionalisation capitaliste: Témoignages de Fos et du Languedoc.* Paris: Anthropos, 1973.

Ehrmann, Henry W. "The French Trade Associations and the Ratification of the Schuman Plan." *World Politics* 6, no. 4 (July 1954): 453–81.

———. *Organized Business in France.* Princeton: Princeton University Press, 1957.

Engerand, C., and M. Perricaudet. "Bilan et perspectives de la sidérurgie française à l'orée du VIIe Plan." *Annales des Mines*, June 1976, 9–16.

Erbès-Seguin, Sabine. *Syndicats et relations de travail dans la vie économique française.* Lille: Presses Universitaires de Lille, 1985.

Erdemli, Hayrettin, Bernard Réal, Pierre Judet, and C. Palloix. *L'Internationalisation dans la sidérurgie.* Grenoble: Université des sciences sociales, Institut de Recherches Économique et de Planification, 1974.

Estrin, Saul, and Peter Holmes. *French Planning in Theory and Practice.* London: Allen & Unwin, 1983.

Eyraud, François, and Robert Tchobanian. "The Auroux Reforms and Company Level Industrial Relations in France." *British Journal of Industrial Relations* 23, no. 2 (July 1985): 241–59.

Fabbra, Daniel."Compétences: Un challenge en acier pour Sollac Florange." *Education permanente* 105 (December 1990).

Fahy, Jean-Michel. *Le chômage en France*, 4th ed. Paris: PUF, Que sais-je?, 1989.

Flanagan, Robert J. David W. Soskice, and Lloyd Ulman. *Unionism, Economic Stabilization, and Incomes Policies: European Experience.* Washington, D.C.: Brookings, 1983.

Fourastié, Jean. *Les trente glorieuses ou la révolution invisible.* Paris: Fayard, 1979.

Freysennet, Michel. *La sidérurgie française, 1945–1979: L'Histoire d'une faillite, Les solutions qui s'affrontent.* Paris: Savelli, 1979.

Freysennet, Michel, and Françoise Imbert. *La Centralisation de capital dans la sidérurgie, 1945–1975.* Paris: Centre de Sociologie Urbaine, 1975.

Freysennet, Michel, and Catherine Omnès. *La Crise de la sidérurgie française.* Paris: Hatier, 1982.

Freyssinet, Jacques. *Le chômage.* Paris: La Découverte, 1987.

Friedberg, Erhard. "Administration et entreprises." In Michel Crozier, ed., *Où va l'administration française?* Paris: Éditions d'Organisation, 1974, 101–40.

Fritsch, Pierre. *Les Wendel, rois de l'acier français.* Paris: Robert Laffont, 1976.

Gallie, Duncan. *Social Inequality and Class Radicalism in France and Britain.* Cambridge: Cambridge University Press, 1983.

———. "*Les lois Auroux*: The Reform of French Industrial Relations?" In Howard Machin and Vincent Wright, eds., *Economic Policy and Policy-Making Under the Mitterrand Presidency, 1981–1984.* New York: St. Martin's, 1985, 205–21.

Gaudemet, Yves. "Les contrats de solidarité (Aspects de droit public)." *Droit social* 45, no. 4 (April 1982): 335–44.

Gaumé, Gérard. "L'Esquive du réel." *Droit Social* 47, no. 4 (April 1984): 249–53.

Gauron, André. *Histoire économique et sociale de la Cinquième République: Le temps des modernistes*, Vol. 1. Paris: La Découverte, 1983.

———. *Histoire économique et sociale de la Cinquième République: Années de rêves, années de crises (1970–1981)*, Vol. 2. Paris: La Découverte/Maspero, 1988.

Géhin, Jean-Paul."Les institutions de formation continue: Diversité des interventions patronales." *Formation emploi* 11 (July–September 1985): 48–61.

Gerschenkron, Alexander. *Economic Backwardness in Historical Perspective.* Cambridge: Harvard University Press, 1962.

Gillingham, John. *Coal, Steel, and the Rebirth of Europe, 1945–1955: The Germans and French from Ruhr Conflict to Economic Community.* Cambridge: Cambridge University Press, 1991.

Glyn, Andrew, Alan Hughes, Alain Lipietz, and Ajit Singh. "The Rise and Fall of the Golden Age." In Stephen A. Marglin and Juliet B. Schor, eds. *The Golden Age of Capitalism.* Oxford: Clarendon Press, 1990.

Gonod, Pierre. *Problèmes et méthodes de mesure de la productivité dans les industries de la Communauté.* Paris: Dalloz and Sirey, 1961.

Gonot, J. "La Concentration sidérurgique mosellane: Perspectives économiques." Mémoire de stage. École Nationale d'Administration, December 1950.

Gourc, Geneviève, and Jacques Gourc. "La restructuration de la sidérurgie dans les usines performantes: la Solmer à Fos-sur-Mer." *Critiques de l'Économie Politique*, nos. 15–16 (1981): 135–76.

Gourdet, Geneviève. "Une forme spécifique de financement de la sidérurgie." In Boy et al., eds., *Sidérurgie et droit de crise*, 75–103.

Granger, Georges. "Produire: Pour qui? Pourquoi?" *Dialectiques* 28 (Autumn 1979): 99–108.

Gravière, Isabelle. "Le plan français: choisir entre l'Est et l'Ouest," *Le Nouveau Journal*, December 9, 1982.

Green, Diana. *Managing Industrial Change? French Policies to Promote Industrial Adjustment*. London: HMSO, 1981.

Groux, Guy, and René Mouriaux. *La C.F.D.T.* Paris: Economica, 1989.

Groux, Guy, and René Mouriaux. *La C.G.T.: Crises et alternatives*. Paris: Economica, 1992.

Groux, Guy, and Cathérine Lévy. "Mobilisation collective et productivité économique: le cas des cercles de qualité dans la sidérurgie." *Revue française de socioloqie* 26, no. 1 (January–March 1985): 70–95.

Groux, Guy, and Cathérine Lévy. "Savoir, travail et production: La sidérurgie à Thionville," *Travail* 14 (October 1987).

Grunert, Thomas. "Decision-Making Processes in the Crisis Policy of the EEC: Neocorporatist or Integrationist Tendencies?" In Mény and Wright, eds., *The Politics of Steel*, 222–307.

Guienne, Raymond, and André Pierrard. *Denain: Un crime signé Usinor*. Condé-sur-Escaut: G. Blondel. 1979.

Hall, Peter A. *Governing the Economy: The Politics of State Intervention in Britain and France*. New York: Oxford University Press, 1986.

Hamon, Hervé, and Patrick Rotman. *La deuxième gauche: Histoire intellectuelle et politique de la CFDT*. Paris: Éditions Ramsay, 1982.

Hardy-Hémery, Odette. "Permanences et mouvements dans un pôle industriel: Le Valenciennois de septembre 1944 à 1947." *Revue d'histoire de la Deuxième Guerre Mondiale*, April 1976, 83–108.

———. *De la croissance à la désindustrialisation: Un siècle dans le Valenciennois*. Paris: Presses de la Fondation Nationale des Sciences Politiques, 1984.

Harris, André, and Alan de Sédouy. *Les patrons*. Paris: Seuil, 1977.

Hayward, Jack. "Le fonctionnement des Commissions et la préparation du Ve Plan: L'exemple de la Commission de la Main-d'oeuvre." *Revue française de sociologie* 8, no. 4 (October–December 1967): 447–67.

———. "Steel." In Raymond Vernon, ed., *Big Business and the State: Changing Relations in Western Europe*. Cambridge: Harvard University Press, 1974, 229–71.

———. *The State and the Market Economy: Industrial Patriotism and Economic Intervention in France*. New York: New York University Press, 1986.

———. "The Nemesis of Industrial Patriotism: The French Response to the Steel Crisis." In Mény and Wright, eds., *The Politics of Steel*, 502–33.

Hocquet, Jean-Yves. "Les conventions du Fonds national de l'emploi." *Droit social* 48, no. 9–10 (September–October 1985): 651–57.

———. "La reconversion des salariés." *Regards sur l'actualité*, no. 134 (October 1987): 34–46.

Hoffmann, Stanley. "Paradoxes of the French Political Community." In Hoffmann et al., *In Search of France: The Economy, Society, and Political System in the Twentieth Century*. New York: Harper, 1963, 1–117.

———. *Decline and Renewal? France Since the 1930s*. New York: Viking, 1974.

Hoffmann-Martinot, Vincent, and Pierre Sadran. "The Local Implementation of France's National Strategy." In Mény and Wright, eds., *The Politics of Steel*, 534–92.

Hogan, William T. *World Steel in the 1980s: A Case of Survival*. Lexington, MA: D.C. Heath, 1983.

Hollingsworth, J. Rogers, Philippe C. Schmitter, and Wolfgang Streeck. "Capitalism, Sectors, Institutions, and Performance." In Hollingsworth, Schmitter, and Streeck, eds., *Governing Capitalist Economies: Performance and Control of Economic Sectors*. New York: Oxford, 1994.

Holter, Darryl. *The Battle for Coal: Miners and the Politics of Nationalization, 1940–1950*. DeKalb: Northern Illinois University Press, 1992.

Howell, Chris. "The Fetishism of Small Difference: French Socialism Enters the Nineties." *French Politics and Society* 9, no. 1 (1990).

———. *Regulating Labor: The State and Industrial Relations Reform in Postwar France*. Princeton: Princeton University Press, 1992.

Howell, Chris, and Anthony Daley. "Introduction: The Transformation of Political Exchange." *International Journal of Political Economy* 23, no. 4 (Winter 1992–93).

Huc, Michel. "La nouvelle convention de protection sociale de la sidérurgie." *Droit social* 47, no. 11 (November 1984): 640–55.

Huiban, Jean-Pierre. "The Industrial Counterproposal as an Element of Trade Union Strategy." In Kesselman with Groux, eds., *The French Workers' Movement*, 224–37.

International Iron and Steel Institute. *Causes of the Mid-1970s Recession in Steel Demand*. Brussels: IISI, 1980.

———. *Indirect Trade in Steel—1962 to 1979*. Brussels: IISI, 1982.

d'Iribarne, Alain. *La Compétitivité: Défi social, enjeu éducatif*. Paris: Presses du CNRS, 1989.

d'Iribarne, Philippe. "Régulation sociale, vie des entreprises et performance économiques." *Revue économique* 37, no. 3 (May 1986): 429–54.

Jackman, Patrick C. "Unit Labor Cost of Iron and Steel Industries in Five Countries." *Monthly Labor Review* 92, no. 8 (August 1969).

Jacquier, Jean-Paul. "Quelques observations." *Droit Social* 47, no. 4 (April 1984): 254.

Jeandin, Jacques. *Trieux: 79 jours au fond pour la Lorraine*. Paris: Éditions sociales, 1977.

Jenson, Jane, and George Ross. "The Tragedy of the French Left." *New Left Review* 171 (September/October 1988): 5–46.

Jobert, Annette. "La Négociation collective dans les entreprises multi-nationales en Europe." In Guillaume Devin, ed., *Syndicalisme: Dimensions internationales.* Garenne-Colombes: Éditions Européenes Erasme, 1990.

Jobert, Bruno. "Le Ministère de l'Industrie et la cohérence de la politique industrielle." *Revue française de science politique* 23, no. 2 (April 1973), 321–29.

July, Serge. *Les Années Mitterrand: Histoire baroque d'une normalisation inachevée.* Paris: Grasset, 1986.

Kassalow, Everett M. "Crisis in the World Steel Industry: Union-Management Responses in Four Countries." *Labour and Society* 10, no. 3 (September 1985): 345–67.

Katzenstein, Peter J. *Small States in World Markets: Industrial Policy in Europe.* Ithaca: Cornell University Press, 1985.

Keeler, John T. S. "Corporatist Decentralization and Commercial Modernization in France: The Royer Law's Impact on Shopkeepers, Supermarkets and the State." In Philip G. Cerny and Martin A. Schain, eds., *Socialism, the State and Public Policy in France.* New York: Methuen, 1985, 265–91.

———. "Situating France on the Pluralism-Corporatist Continuum: A Critique and Alternative to the Wilson Perspective." *Comparative Politics* 17, no. 2 (January 1985): 229–49.

———. *The Politics of Neo-Corporatism in France.* New York and Oxford: Oxford University Press, 1987.

Kerschen, Nicole and Huguette Reminiac. "Les systèmes de 'pré-retraite': Classification juridique et pratique." *Droit social* 44, no. 2 (February 1982): 175–201.

Kerschen, Nicole, and Anne-Valérie Nenot. "Délégation à l'emploi et négociation des conventions du Fonds national de l'emploi: la pratique des contreparties." *Droit social* 51, no. 1 (January 1989): 17–22.

Kesselman, Mark. "French Labor Confronts Technological Change: The Story of a Reform that Never Was." In Daley, ed., *The Mitterrand Era.*

Kesselman, Mark, with Guy Groux, eds., *The French Workers' Movement: Economic Crisis and Political Change.* London: Allen & Unwin, 1984.

Kitschelt, Herbert P. "Political Opportunity Structures and Political Protest: Anti-Nuclear Movements in Four Democracies." *British Journal of Political Science* 16 (January 1986), pp. 57–85.

Korpi, Walter, and Michael Shalev. "Strikes, Industrial Relations and Class Conflict in Capitalist Societies." *British Journal of Sociology* 30, no. 2 (June 1979).

Kuisel, Richard F. *Capitalism and the State in Modern France: Renovation and Economic Management in the Twentieth Century.* Cambridge: Cambridge University Press, 1981.

———. "L'American Way of Life et les Missions Françaises de Productivité." *Vingtième siècle* 17 (January–March 1988): 21–38.

Labbé, Dominique, and Maurice Croisat. *La fin des syndicats?* Paris: l'Harmattan, 1992.

Lachaise, Gérard. "Les contrats à durée déterminé successifs." *Droit social* 52, nos. 9–10 (September–October 1989).

de Lacoste-Lareymondie, Alain. "La participation des employeurs à la construction de logements." *Droit social* 17, no. 3 (March 1954): 153–60.

Laffont, Paul. "Les entreprises sidérurgiques depuis cinq ans." *Economie et Statistique* 27 (October 1971): 43–51.

Lagrange, François. "Manpower and Employment Policies in France." *Labour and Society* 5, no. 2 (April 1980): 119–21.

Lalanne, Bernard. "Mort d'un Lobby." *L'Expansion* (June 1979): 104–11.

Landes, David S. "French Entrepreneuship and Industrial Growth in the Nineteenth Century." *Journal of Economic History*, May 1949: 45–61.

Landier, Hubert. "La crise du syndicalisme et l'évolution des comportements patronaux." *The Tocqueville Review* 2, nos. 2–3 (Spring–Summer 1980): 110–22.

———. *Demain, quels syndicats?* Paris: Livre de Poche, 1981.

Langlois, Philippe. "La politique des avantages acquis." *Droit social* 49, no. 12 (December 1986): 881–86.

Laroque, Michel. *Politiques sociales dans la France contemporaine: Le social face à la crise*. Paris: Éditions STH, 1984.

Laroque, Pierre, ed., *Les institutions sociales de la France*. Paris: La Documentation Française, 1980.

Lash, Scott. *The Militant Worker: Class and Radicalism in France and America*. Rutherford, N.J.: Fairleigh Dickinson University Press, 1984.

Lash, Scott, and John Urry. *The End of Organized Capitalism*. Madison: University of Wisconsin Press, 1987.

Lefour, Alain. "Wendel-Sidélor: Le prix de la mutation." *Entreprise* 847 (December 3, 1971): 178–89.

Lefranc, Georges. *Le Mouvement syndical en France: De la Libération aux événements de mai-juin 1968*. Paris: PUF, 1969.

Lehmbruch, Gerhard, and Philippe C. Schmitter, eds. *Patterns of Corporatist Policy-Making*. Beverly Hills and London: Sage, 1982.

Lehut, Michel. *La Sidérurgie dans le Valenciennois, 1966–1979*. Master's Thesis. University of Lille III, 1981.

Leuthold, Patrice et al. *L'Entreprise et le syndicalisme*. Paris: CNRS-IRESCO, 1988.

Lévy-Leboyer, Maurice. "Le patronat français a-t-il été malthusien?" *Mouvement social* 88 (July–September 1974): 3–49.

———. "The Large Corporation in Modern France." In Alfred Chandler, ed., *Managerial Hierarchies: Comparative Perspectives on the Rise of the Modern Industrial Enterprise*. Cambridge: Havard University Press, 1980, 117–60.

Lindberg, Leon N., John L. Campbell, and J. Rogers Hollingsworth, eds. *Governance of the American Economy.* Cambridge: Cambridge University Press, 1991.

Lipietz, Alain. "Governing the Economy in the Face of International Challenge: From National Developmentalism to National Crisis." In James F. Hollifield and George Ross, eds., *Searching for the New France.* New York: Routledge, 1991.

Lister, Louis. *Europe's Coal and Steel Community: An Experiment in Economic Union.* New York: Twentieth Century Fund, 1960.

Loriaux, Michael. *France After Hegemony: International Change and Financial Reform.* Ithaca: Cornell University Press, 1991.

"Lorraine." *Espace et luttes*, Special Issue, nos. 3–4 (1980).

Lorraine, Coeur d'Acier! Dir. Alban Poirier and Jean Serres, 1981.

Lorwin, Val R. *The French Labor Movement.* Cambridge: Harvard University Press, 1954.

Loubéjac, François. "Sur la suppression de l'autorisation de licenciement." *Droit social* 49, no. 3 (March 1986): 213–19.

Lynch, Frances. "The Role of Jean Monnet in Setting up the European Coal and Steel Community." In Schwabe, ed., *The Beginnings of the Schuman-Plan*, 117–29.

Lyon-Caen, Antoine. "Le comité d'entreprise et la restructuration de l'entreprise." *Droit social* 42, no. 4 (April 1979): 23–34.

———. "Analyse juridique des congés de conversion." *Droit social* 48, nos. 9–10 (September–October 1985): 660–671.

Lyon-Caen, Gérard. "Droit syndical et mouvement syndical." *Droit social* 47, no. 1 (January 1984): 5–14.

———. "La bataille truquée de la flexibilité." *Droit social* 48, no. 12 (December 1985): 801–10.

Lyon-Caen, Gérard, and M.-C. Bonnetete. "La réforme du licenciement à travers la loi du 13 juillet 1973." *Droit social* 36, no. 11 (November 1973): 493–507.

Maire, Edmond. "Du changement imposé aux adaptation négociées." *Droit social* 49, no. 12 (December 1986): 834–36.

Maire, Edmond, Charles Piaget, et al. *Lip 73.* Paris: Seuil, 1973.

Malézieux, Jacques. "Crise et restructuration de la sidérurgie française: Le groupe Usinor." *L'Espace Géographique* 3 (July–September 1980): 183–96.

———. *Les centres sidérurgiques des rivages de la Mer du Nord et leur influence sur l'organisation de l'espace.* Paris: Cujas, 1981.

"Le Management participatif à l'épreuve." *Travail*, Special Issue, no. 24 (Winter 1991–1992).

Marioni, Pierre. "Les congés de conversion." *Dossiers statistiques du travail et de l'emploi*, 34–35 (October 1987): 107–14.

———. "Les dispositifs de conversion." *Dossiers statistiques du travail et de l'emploi*, nos. 43–44 (October 1988): 66–67.

Martin, Dominique. "Les systèmes de négociation et de représentation dans l'entreprise." *Droit social* 39, no. 3 (March 1976): 92–101.

Martin, Dominique, ed. *Participation et changement social dans l'entreprise.* Paris: L'Harmattan, 1989.

Martin, Roger. *Patron de droit divin.* Paris: Gallimard, 1984.

Masson, Jean-Louis. *L'Industrie lorraine du fer . . . ou comment tuer la poule aux oeufs d'or.* Paris: Cujas, 1977.

———. *Demain l'acier français.* Paris: Cujas, 1986.

Maurice, Marc, and François Sellier. "Societal Analysis of Industrial Relations: A Comparison Between France and West Germany." *British Journal of Industrial Relations* 17, no. 3 (November 1979): 322–36.

Maurice, Marc, and François Sellier, and Jean-Jacques Silvestre. *Politique d'éducation et organisation industrielle en France et en Allemagne: Essai d'analyse sociétale.* Paris: PUF, 1982.

McArthur, John H., and Bruce R. Scott. *Industrial Planning in France.* Boston: Harvard Graduate School of Business, 1969.

McCormick, Janice. "Gaullism and Collective Bargaining: The Effect of the Fifth Republic on French Industrial Relations." In Andrews and Hoffmann, eds., *The Fifth Republic at Twenty.*

Mény, Yves, and Vincent Wright, eds. *La crise de la sidérurgie européene, 1974–1984.* Paris: Presses Universitaires de France, 1985.

Mény, Yves, and Vincent Wright, eds. *The Politics of Steel: Western Europe and the Steel Industry in the Crisis Years (1974–1984).* Berlin and New York: de Gruyter, 1987.

Messerlin, Patrick. "La Sidérurgie européene face à la crise mondiale." In Mény and Wright, eds., *La crise*, 143–76.

———. "The European Iron and Steel Industry and the World Crisis." In Mény and Wright, eds., *The Politics of Steel.*

Messerlin, Patrick, and Christopher Saunders. "Steel: Too Much Investment Too Late." In Geoffrey Shephard, François Duchêne and Christopher Saunders, eds., *Europe's Industries: Public and Private Strategies for Change.* London: Frances Pinter, 1983, 52–81.

Meynaud, Jean. *Les groupes de pression en France.* Paris: A. Colin, 1959.

Michel, Andrée. "L'Immigration algérienne en Moselle." *Annales de Géographie* 65, no. 5 (September–October 1956): 341–61.

Millot, Michèle, and Jean-Pol Roulleau. *L'Entreprise et les strategies syndicales.* Paris: Éditions d'Organisation, 1988.

Ministère du Travail, de l'Emploi et de la Formation Professionnelle. *Dossiers statistiques du travail et de l'emploi* 68 (April 1991).

Ministère de la Recherche et de l'Industrie, ed. *Une politique industrielle pour la France: Actes des journées de travail des 15 et 16 novembre 1982*. Paris: Documentation Française, 1983.

Mioche, Philippe. "Le démarrage du Plan Monnet: Comment une entreprise conjoncturelle est devenue une institution politique." *Revue d'histoire moderne et contemporaine* 31, no. 3 (July–September 1984): 398–416.

———. *Le Plan Monnet: Genèse et élaboration, 1941–1947*. Paris: Publications de la Sorbonne, 1987.

———. "Le patronat de la sidérurgie française et le Plan Schuman en 1950–1952: Les apparences d'un combat et la réalité d'une mutation." In Schwabe, ed., *The Beginnings of the Schuman-Plan*, 305–18.

———. *Jacques Ferry et la sidérurgie française depuis la Seconde Guerre Mondial*. Aix-en-Provence: Publications de l'Université de Provence, 1993.

Mioche, Philippe, and Jacques Roux. *Henri Malcor: Un héritier des maîtres de forge*. Paris: Editions du CNRS, 1988.

Monnet, Jean. *Memoirs*. Trans. Richard Mayne. London: Collins, 1978.

de Montlibert, Christian. "Chômage et licenciements. La crise de la Lorraine sidérurgique: Longwy." *Annales* 39, no. 5 (September–October 1984): 1044–68.

———. *Crise économique et conflits sociaux*. Paris: L'Harmattan, 1989.

Moravcsik, Andrew. "Negotiating the Single Act: National Interests and Conventional Statecraft in the European Community." In Robert O. Keohane and Stanley Hoffmann, eds., *The New European Community: Decisionmaking and Institutional Change*. Boulder: Westview, 1991.

Morin, François. *La Structure financière du capitalisme français*. Paris: Calmann-Lévy, 1974.

Morville, Pierre. *Les nouvelles politiques sociales du patronat*. Paris: La Découverte, 1985.

Moss, Bernard H. *The Origins of the French Labor Movement, 1830–1914: The Socialism of Skilled Workers*. Berkeley and Los Angeles: University of California Press, 1976.

———. "After the Auroux Laws: Employers, Industrial Relations and the Right in France." *West European Politics* 11, no. 1 (January 1988): 68–80.

Mourgues, Bernard. "La convention de protection sociale de la sidérurgie." *Droit social* 41, no. 2 (February 1978): 157–69.

Mouriaux, René. *La CGT*. Paris: Seuil, 1982.

———. *Les syndicats dans la société française*. Paris: Presses de la Fondation Nationale des Sciences Politiques, 1983.

———. "The CFDT: From the Union of Popular Forces to the Success of Social Change." In Kesselman with Groux, eds. *The French Workers' Movement*, 75–92.

Moynot, Jean-Louis. "Le mouvement syndical en mouvement." *Dialectiques* 28 (Autumn 1979): 36–55.

———. *Au milieu du gué: CGT, syndicalisme et démocratie de masse.* Paris: PUF, 1982.

Neuschwander, Claude. *Patron, mais . . . : Entretiens avec Bernard Guetta.* Paris: Seuil, 1975.

Noblecourt, Michel. "France: Les rude leçons du 'spontanéisme' social." *Revue française des affaires sociales* 41, no. 1 (January–March 1987): 57–64.

———. *Les syndicats en questions.* Paris: Éditions Ouvrières, 1990.

Noël, Alain. "Accumulation, Regulation, and Social Change: An Essay on French Political Economy." *International Organization* 41, no. 2 (Spring 1987), 303–33.

Noiriel, Gérard. *Longwy: Immigrés et prolétaires, 1880–1980.* Paris: PUF, 1984.

Noiriel, Gérard, with Benaceur Azzaoui. *Vivre et lutte à Longwy.* Paris: Maspero, 1980.

Organization for Economic Cooperation and Development. *Steel in the 80s.* Paris: OECD, 1980.

Orum, Anthony M., Joe R. Feagin, and Gideon Sjoberg. "Introduction: The Nature of Case Studies." In Feagin, Orum, and Sjoberg, eds., *A Case for the Case Study.* Chapel Hill: University of North Carolina Press, 1991.

Outin, Jean-Luc, and Rachel Silvera. "Les entreprises dans la reconversion: L'exemple du Creusot." *Économie et Humanisme* 304 (November–December 1988).

Padioleau, Jean G. *Quand la France s'enferre: La politique sidérurgique de la France depuis 1945.* Paris: PUF, 1981.

Paillard, Bernard with Claude Fischler. *La damnation de Fos.* Paris: Seuil, 1981.

Paxton, Robert O. *Vichy France: Old Guard and New Order, 1940–1944.* New York: Columbia University Press, 1972.

Peco, Franco. "How Steelmaking Enterprises Can Become Internationally Competititve." In OECD, *Steel in the 80s,* 127–39.

Peyrefitte, Alain. *Le mal français.* Paris: Plon, 1977.

Piaget, Charles. *Lip: Charles Piaget et les LIP racontent.* Paris: Stock, 1973.

Piore, Michael J., and Charles F. Sabel. *The Second Industrial Divide: Possibilities for Prosperity.* New York: Basic, 1984.

Polsby, Nelson W. *Community Power and Political Theory: A Further Look at Problems of Evidence and Inference,* 2d ed. New Haven: Yale University Press, 1980.

Poulantzas, Nicos. *State, Power, Socialism.* London: Verso, 1980.

Prêcheur, Claude. *La Lorraine sidérurgique.* Paris: S.A.B.R.I., 1959.

———. *La sidérurgie française.* Paris: Armand Colin, 1963.

Reid, Donald. "The Limits of Paternalism: Immigrant Coal Miners' Communities in France, 1919–45." *European History Quarterly* 15, no. 1 (January 1985): 99–118.

Reiss, Elisabeth. "Interview à la Sollac." *Esprit* 348 (April 1966): 686–94.
Reitel, François. *La Lorraine.* Paris: PUF-Que sais-je?, 1982.
Reynaud, Jean-Daniel. "La convention sociale de la sidérurgie lorraine." *Droit social* 32, no. 4 (April 1969): 219–27.
———. *Les syndicats en France*, 2 Vols. Paris: Seuil, 1975.
———. "Trade Unions and Political Parties in France: Some Recent Trends." *Industrial and Labor Relations Review* 28, no. 2 (January 1975): 208–25.
———. *Les syndicats, les patrons et l'état: Tendances de la negociation collective en France.* Paris: Editions Ouvrières, 1978.
———. "Conflits du travail, classes sociales et contrôle social." *Sociologie du travail* 22, no. 1 (January–March 1980): 97–115.
———. "France: Elitist Society Inhibits Articulated Bargaining." In Solomon Barkin, ed., *Worker Militancy and Its Consequences: The Changing Climate of Western Industrial Relations*, 2d ed. New York: Praeger, 1983, 279–311.
———. "La négociation des nouvelles technologies: Une transformation des règles du jeu?" *Revue française de science politique* 38, no. 1 (February 1988): 5–22.
Rhodes, Martin, and Vincent Wright. "The European Steel Unions and the Steel Crisis, 1974–84: A Study in the Demise of Traditional Unionism." *British Journal of Political Science* 18, no. 2 (April 1988): 171–95.
Rideau, Emile. *Essor et problèmes d'une région française: Houillères et sidérurgie de Moselle.* Paris: Editions Ouvrières, 1956.
Risse, Jean. *Leur silence est parole: La mémoire d'un prêtre-ouvrier* (Paris: Editions Ouvrières, 1991).
Rocard, Michel, and the Clubs Convaincre. *Construire le progrès social.* Paris: Syros, 1990.
Rognant, Jean-Yves, Camille Romain, and François Rosso. *Où va la sidérurgie?* Paris: Syros, 1977.
Rosanvallon, Pierre. *La question syndicale.* Paris: Calamann-Lévy, 1988.
———. "The Development of Keynesianism in France," in Peter A. Hall, ed., *The Political Power of Economic Ideas: Keynesianism Across Nations.* Princeton: Princeton University Press, 1989, 171–93.
Rose, José. *En quête d'emploi: Formation, chômage, emploi.* Paris: Economica, 1984.
Rose, Michael. "Economic Nationalism and the Unions: The Decline of the *Solution Franco-Française.*" In Jolyon Howorth and George Ross, eds., *Contemporary France*, Vol. 2. London: Pinter, 1988.
Ross, George. "The Confédération Générale du Travail in Eurocommunism." *Politics and Society* 9, no. 1 (1979): 33–60.
———. "French Labor and Economic Change." In Cohen and Gourevitch, eds., *France in a Troubled World Economy*, 151–179.
———. "French Trade Unions Face the 1980s: The CGT and the CFDT in the

Strategic Conflicts and Economic Crisis of Contemporary France." In *Political Power and Social Theory*, vol. 3 (1982), 53–76.

———. "The Perils of Politics: French Unions and the Crisis of the 1970s." In Peter Lange, George Ross, and Maurizio Vanicelli, *Unions, Change and Crisis: French and Italian Union Strategy and the Political Economy, 1945–1980*. London and Boston: George Allen & Unwin, 1982, 13–93.

———. *Workers and Communists in France: From Popular Front to Eurocommunism*. Berkeley and Los Angeles: University of California Press, 1982.

———. "Adieu vieilles idées: The Middle Strata and Decline of Resistance-Liberation Left Discourse in France." In Jolyon Howorth and Ross, eds., *Contemporary France*, Vol. 1. London: Pinter, 1987.

———. "French Intellectuals from Sartre to Soft Ideology." In Charles C. Lemert, ed., *Intellectuals and Politics: Social theory in a Changing World*. Newbury Park: Sage, 1991.

Roth, François. "Les milieux sidérurgiques et l'annonce du Plan Schuman." In Schwabe, ed., *The Beginnings of the Schuman-Plan*, 367–80.

Sabel, Charles. *Work and Politics: The Division of Labor in Society*. Cambridge: Cambridge University Press, 1982.

Sainsaulieu, Renaud, ed. *L'Entreprise: Une affaire de société*. Paris: Presses de la Fondation Nationale des Sciences Politiques, 1990.

Saint-Jours, Yves. "France." In Peter A. Köhler and Hans F. Zacher, eds., *The Evolution of Social Insurance, 1991–1981: Studies of Germany, France, Great Britain, Austria and Switzerland*. New York: St. Martin's, 1982, 93–149.

Saint-Marc, Philippe. *La France dans la C.E.C.A.: Une expérience de planifications multiples du charbon et de l'acier*. Cahier de la Fondation Nationale des Science Politiques, no. 114. Paris: Armand Colin, 1961.

Salah, M. M. Mohamed. "Aspects sociaux de la crise sidérurgique." In Boy et al., ed., *Sidérurgie et droit de crise*, 135–70.

Salisbury, Robert. *Interests and Institutions: Substance and Structure in American Politics*. Pittsburgh: University of Pittsburgh Press, 1992.

Sandholtz, Wayne, and John Zysman. "1992: Recasting the European Bargain." *World Politics* 42, no. 1 (October 1989).

Savage, Dean. *Founders, Heirs, and Managers: French Industrial Leadership in Transition*. Beverly Hills and London: Sage, 1979.

Savouillan, Charles. "Les salaires des ouvriers dans les industries de la Communauté." *Droit Social* 21, no. 5 (May 1958): 271–83.

———. "Les salaires des ouvriers dans les industries de la Communauté." *Droit Social* 21, no. 6 (June 1958): 347–53.

———. "L'harmonisation des salaires et des conditions de travail dans les industries de la C.E.C.A." *Droit Social* 23, nos. 9–10 (September–October 1960): 472–86.

Schain, Martin. "Corporatism and Industrial Relations in France." In Philip G. Cerny and Schain, eds., *French Politics and Public Policy*. New York: St. Martin's, 1980, 191–217.

———. "The Dynamics of Labor Policy in France: Industrial Relations and the French Trade Union Movement." *The Tocqueville Review* 2, no. 1 (Winter 1980): 77–115.

Schattschneider, E.E. *The Semisoveriegn People: A Realist's View of Democracy in America*. New York: Holt, Rinehart and Winston, 1960.

Schmidt, Vivien. *Democratizing France: The Political and Administrative History of Decentralization*. Cambridge: Cambridge University Press, 1991.

———. "Through 'More State' and 'Less State,' the French State Goes On: Patterns of Industrial Policy-Making in France 1981–1991." Paper presented to the Annual Meeting of the American Political Science Association. Washington, D.C., 1991.

———. "An End to French Economic Exceptionalism? The Transformation of Business under Mitterrand." In Daley, ed., *The Mitterrand Era*.

Schmitter, Philippe C., and Gerhard Lehmbruch, eds. *Trends Toward Corporatist Intermediation*. Beverly Hills and London: Sage, 1979.

Schonfield, William. *Obedience and Revolt: French Behavior Toward Authority*. Beverly Hills and London: Sage, 1976.

Schwabe, Klaus ed. *The Beginnings of the Schuman-Plan*. Baden-Baden: Nomos Verlag, 1988.

Schwob, René. "Souvenirs de trois décade d'activité syndicale." Metz: Syndicats Force Ouvrière de la Moselle, 1957.

———. "Un Futur centenaire, Le syndicalisme français." Metz: CGT-FO, 1981.

Segrestin, Dénis. *Le Phénomène corporatiste: Essai sur l'avenir des systèmes professionnels fermés en France*. Paris: Fayard, 1985.

Sellier, François. "La cohésion patronale dans les négociations sociales: L'exemple de l'union des industries métallurgiques et minières." *Droit social* 23, nos. 7–8 (July–August 1960): 411–16.

———. *Stratégie de la lutte sociale: France, 1936–1960*. Paris: Editions Ouvrières, 1961.

———. "L'évolution des négociations collectives dans la sidérurgie et la métallurgie en France (1950–1969)." *Droit social* 33, nos. 9–10 (September–October 1970): 431–49.

———. *La confrontation sociale en France, 1936–1981*. Paris: PUF, 1984.

———. "Aménagement du temps de travail, articulation entre niveaux de négociation et conflits entre syndicats." *Droit social* 49, no. 11 (November 1986): 773–78.

Sheahan, John. *Promotion and Control of Industry in Postwar France*. Cambridge: Harvard University Press, 1963.

Shonfield, Andrew. *Modern Capitalism: The Changing Balance of Public and Private Power.* London and New York: Oxford University Press, 1965, 1969.

Shorter, Edward, and Charles Tilly. *Strikes in France, 1830–1968.* Cambridge: Cambridge University Press, 1974.

Simon, S. "Histoire d'une grève: La grève de Micheville (28 janvier–16 mars 1951)." *Droit social* 14, no. 6 (June 1951): 390–96; no. 7 (July–August 1951): 466–70; no. 8 (September–October 1951): 535–39.

Smith, W. Rand. *Crisis in the French Labour Movement: A Grassroots' Perspective.* New York: St. Martin's, 1987.

———. "The Mitterrand Government and Industrial Crisis: Restructuring in the Steel and Automobile Industries." In Daley, ed., *The Mitterrand Era.*

Société pour la conversion et le développement industriels. "Études des perspectives de développement économique du bassin ferrifère et sidérurgique lorrain: Rapport de synthèse." Paris: Sodic, 1965.

Soisson, Jean-Pierre, Jean-François de Martel, and Bruno Rémond. *L'Enjeu de la formation professionnelle.* Paris: Fayard, 1986.

Spierenburg, Dirk, and Raymond Poidevin. *Histoire de la Haute Autorité de la Communauté du Charbon et de l'Acier: Un expérience supranationale* (Brussels: Bruylont, 1993).

Steines, George. *Le fer dans la peau.* Metz: Éditions Serpenoise, 1977.

Stoffäes, Christian. "Le dysfonctionnement du système acier." *Revue d'économie industrielle* 8, no. 2 (1979): 114–23.

———. "French Industrial Strategy in Sunrise Sectors." In Alexis Jacquemin, ed., *European Industry: Public Policy and Corporate Strategy.* Oxford: Clarendon Press, 1984, 274–92.

Stoffäes, Christian, and Pierre Gadonneix. "Steel and the State in France." *Annalen der Gemeinwirtschaft* 49, no. 4 (October–December 1980): 393–422.

Stoleru, Lionel. *L'Impératif industriel* (Paris: Seuil, 1969).

Stora, Benjamin. *Crise, puissance, perspectives de la sidérurgie mondiale.* Paris: Économica, 1979.

———. "The World Crisis in the Steel Industry: Outlook for Europe." *Annalen der Gemeinwirtschaft* 49, no. 4 (October–December 1980): 375–91.

Streeck, Wolfgang. *Social Institutions and Economic Performance: Industrial Relations in Advanced Capitalist Economies.* Beverly Hills and London: Sage, 1992.

———. "Industrial Relations in Western Europe: Towards Neo-voluntarism?" Unpublished manuscript, 1992.

Suleiman, Ezra N. *Politics, Power, and Bureaucracy in France: The Administrative Elite.* Princeton: Princeton University Press, 1974.

———. *Elites in French Society: The Politics of Survival.* Princeton: Princeton University Press, 1978.

———. "From Right to Left: Bureaucracy and Politics in France." In Suleiman, ed., *Bureaucrats and Policy Making: A Comparative Overview*. New York: Holmes & Meier, 1984.

Swann, Dennis. *Competition and Industrial Policy in the European Community*. London and New York: Methuen, 1983.

Tanguy, Lucie. *Le capital, les travailleurs et l'école: L'exemple de la Lorraine sidérurgique*. Paris: Maspero, 1976.

Tarrow, Sidney. "Regional Policy, Ideology, and Peripheral Defense: The Case of Fos-sur-Mer." In Tarrow, Peter J. Katzenstein, and Luigi Graziano, eds., *Territorial Politics in Industrial Nations*. New York: Praeger, 1978, 97–122.

Taylor, Charles Lewis, and David A. Jodice. *World Handbook of Political and Social Indicators*, Vol 2: *Political Protest and Government Change*. New Haven: Yale University Press, 1983.

Telesis. *Competing for Prosperity: Business Strategies and Industrial Policies in Modern France*. London: Policy Studies Institute, 1986.

Teyssié, Bernard. "La modification du contrat de travail, instrument de gestion de l'entreprise." *Droit social* 49, no. 12 (December 1986): 852–62.

Thibaud, Paul. "Le communisme municipal." *Esprit* 353 (October 1966): 413–22.

Thouvenin, Monique. "De Wendel-Sidélor à Sacilor-Sollac: 1968–1980. 13 Années de mutations difficiles en Lorraine." *Revue géopraphique de l'Est*, nos. 1–2 (January–June 1981): 37–63.

Tilly, Charles. "Introduction." In Louise A. Tilly and Charles Tilly, eds., *Class Conflict and Collective Action*. Beverly Hills and London: Sage, 1981, 13–25.

———. *The Contentious French*. Cambridge: Harvard University Press, 1986.

Tixier, Pierre-Eric. "Management participatif et syndicalisme." *Sociologie du travail* 28, no. 3 (1986): 353–72.

de Tocqueville, Alexis. *The Old Régime and the French Revolution*. Trans. Stuart Gilbert. New York: Anchor, 1955.

Touraine, Alain. "Unionism as a Social Movement." In Seymour Martin Lipset, ed., *Unions in Transition: Entering the Second Century*. San Francisco: Institute for Contemporary Studies, 1986, 151–73.

Touraine, Alain, Michel Wieviorka, and François Dubet. *Le mouvement ouvrier*. Paris: Fayard, 1984.

Tsoukalis, Loukas, and Robert Strauss. "Community Policies on Steel 1974–1984: A Case of Collective Management." In Mény and Wright, eds., *The Politics of Steel*, 186–221.

Tsoukalis, Loukas, and Robert Strauss. "La politique sidérurgique des communautés: Un cas de gestion collégiale?" In Mény and Wright, eds., *La crise*, 177–218.

Turin, Michel. "Qui contrôlera Creusot Loire?" *Entreprise*, no. 995 (October 4, 1974): 76–79.

U.S. Congress, Office of Technology Assessment. *Technology and Steel Industry Competitiveness.* Washington, D.C.: USGPO, 1980.

Van Lerberghe, Rose-Marie. "La pré-retraite progressive." *Droit social* 45, no. 6 (June 1982): 500–02.

———. "Le nouveau dispositif des congés de conversion." *Droit social* 48, nos. 9–10 (September–October 1985): 672–82.

Vennin, Bruno. "Conversion industrielle: L'État et les autres." Saint-Etienne: CRESAL, 1988.

Verdier, Jean-Maurice. "Sur le nouveau droit de licenciement." *Droit Social* 39, no. 2 (February 1976): 1–18.

Vergnaud, Daniel. "Sidérurgie: Pour une production utile et efficace." *Économie et politique* 61 (May 1982): 35–37.

Vignaux, Paul. *De la CFTC à la CFDT: Syndicalisme et socialisme, "Reconstruction" (1946–1972).* Paris: Éditions ouvrières, 1980.

Villeval, Marie-Claire. "Crise de la sidérurgie et revonversion des salariés." *Les Annales de la recherche urbaine*, nos. 35–36 (November 1987): 21–30.

Villeval, Marie-Claire, ed. *Mutations industrielles et reconversion des salariés.* Paris: l'Harmattan, 1992.

Villeval, Marie-Claire, and Philippe Méhaut. "Les congés de Formation-Conversion et la gestion des mobilités: Une étude de das dans la sidérurgie. *Formation Emploi* 16 (October–December 1986): 59–69.

Vivre à Longwy. Dir. Bernard Gesbert and Jean-Claude Durmeyer. February 1979.

Walkowitz, Daniel J. *Worker City, Company Town: Iron and Cotton-Worker Protest in Troy and Cohoes, New York, 1855–84.* Urbana and Chicago: University of Illinois Press, 1978.

Wall, Irwin W. "Jean Monnet, the United State and the French Economic Plan." In Douglas Brinkley and Clifford Hackett, eds., *Jean Monnet: The Path to European Unity.* New York: St. Martin's, 1991.

Warren, Kenneth. *World Steel: An Economic Geography.* New York: Crane, Russak, 1975.

Weber, Henri. *Le parti des patrons: Le CNPF (1946–1986).* Paris: Seuil, 1986.

Weiss, Dimitri, et al. *La fonction Ressources humaines.* Paris: Éditions d'Organisation, 1988.

Wienert, Helmut. "World Tends in Steel Consumption and Production to 1990." In OECD, *Steel in the 80s*, 83–96.

Wilson, Frank L. *The French Democratic Left, 1963–1969: Toward a Modern Party System.* Stanford: Stanford University Press, 1971.

———. "Alternative Models of Interest Intermediation: The Case of France." *British Journal of Political Science* 12, no. 2 (April 1982), 173–200.

———. "French Interest Group Politics: Pluralist or Neocorporatist?" *American Political Science Review* 77, no. 4 (1983), 895–910.

———. "Les groupes d'intérêt sous la Cinquième République." *Revue française de science politique* 33, no. 2 (April 1983), 220–54.

Wittmann, Michel, and Claude Thouvenot. *La mutation de la sidérurgie: Vers une nouvelle géographie de l'acier.* Paris: Masson et Cie., 1972.

Zarifian, Philippe. "Restructurer or not?" *Dialectiques* 28 (Autumn 1979): 109–18.

———. "Tactiques de lutte dans la sidérurgie." *La nouvelle critique* 123 (April 1979): 57–62.

———. Qualification collective et automatisation: Le cas de la sidérurgie." *Formation emploi* 1 (January–March 1983).

———. "La culture syndicale face à la nécessité de propositions industrielles: Défis, mouvements, portée." *Critiques de l'économie politique*, nos. 23–24 (April–September 1983): 263–84.

———. "La Politique industrielle dans la sidérurgie française de 1977 à 1983: Le jeu de l'echec," *Cahier du Centre de Recherche dur les Mutations des Sociétés Industrielles* 9 (August–September 1984).

———. "Le syndicalisme face à l'entreprise industrielle." *Sociologie du travail* 27, no. 3 (1985): 336–43.

———. Du taylorisme au systemisme: Une nouvelle approche de la qualification dans l'industrie." G.I.P. Mutations Industrielles, no. 8, July 15, 1987.

Zysman, John. *Political Strategies for Industrial Order: State, Market, and Industry in France.* Berkeley and Los Angeles: University of California Press, 1977.

———. "The French State in the International Economy." In Peter J. Katzenstein, ed., *Between Power and Plenty: Foreign Economic Policies of Advanced Industrial States.* Madison: University of Wisconsin Press, 1978, 255–93.

———. "The Interventionist Temptation." In Andrews and Hoffmann, eds., *The Fifth Republic at Twenty,* 252–70.

———. *Governments, Markets, and Growth: Financial Systems and the Politics of Industrial Change.* Ithaca: Cornell University Press, 1983.

———. "Inflation and the the Politics of Supply." In Leon Lindberg and Charles S. Maier, eds., *The Politics of Inflation.* Washington, D.C.: Brookings, 1985.

Index